Professional
Ubuntu® Mobile Development

Professional
Ubuntu® Mobile Development

Ian Lawrence
Rodrigo Cesar Lopes Belem

Wiley Publishing, Inc.

Professional Ubuntu® Mobile Development

Published by
Wiley Publishing, Inc.
10475 Crosspoint Boulevard
Indianapolis, IN 46256
www.wiley.com

Copyright © 2010 by Wiley Publishing, Inc., Indianapolis, Indiana

Published simultaneously in Canada

ISBN: 978-0-470-43676-9

Manufactured in the United States of America

10 9 8 7 6 5 4 3 2 1

No part of this publication may be reproduced, stored in a retrieval system or transmitted in any form or by any means, electronic, mechanical, photocopying, recording, scanning or otherwise, except as permitted under Sections 107 or 108 of the 1976 United States Copyright Act, without either the prior written permission of the Publisher, or authorization through payment of the appropriate per-copy fee to the Copyright Clearance Center, 222 Rosewood Drive, Danvers, MA 01923, (978) 750-8400, fax (978) 646-8600. Requests to the Publisher for permission should be addressed to the Permissions Department, John Wiley & Sons, Inc., 111 River Street, Hoboken, NJ 07030, (201) 748-6011, fax (201) 748-6008, or online at http://www.wiley.com/go/permissions.

Limit of Liability/Disclaimer of Warranty: The publisher and the author make no representations or warranties with respect to the accuracy or completeness of the contents of this work and specifically disclaim all warranties, including without limitation warranties of fitness for a particular purpose. No warranty may be created or extended by sales or promotional materials. The advice and strategies contained herein may not be suitable for every situation. This work is sold with the understanding that the publisher is not engaged in rendering legal, accounting, or other professional services. If professional assistance is required, the services of a competent professional person should be sought. Neither the publisher nor the author shall be liable for damages arising herefrom. The fact that an organization or Web site is referred to in this work as a citation and/or a potential source of further information does not mean that the author or the publisher endorses the information the organization or Web site may provide or recommendations it may make. Further, readers should be aware that Internet Web sites listed in this work may have changed or disappeared between when this work was written and when it is read.

For general information on our other products and services please contact our Customer Care Department within the United States at (877) 762-2974, outside the United States at (317) 572-3993 or fax (317) 572-4002.

Library of Congress Control Number: 2009927341

Trademarks: Wiley, the Wiley logo, Wrox, the Wrox logo, Wrox Programmer to Programmer, and related trade dress are trademarks or registered trademarks of John Wiley & Sons, Inc. and/or its affiliates, in the United States and other countries, and may not be used without written permission. Ubuntu is a registered trademark of Canonical Ltd. All other trademarks are the property of their respective owners. Wiley Publishing, Inc., is not associated with any product or vendor mentioned in this book.

Wiley also publishes its books in a variety of electronic formats. Some content that appears in print may not be available in electronic books.

Introduction xxv

Chapter 1: 1
Chapter 2: 11
Chapter 3: 35
Chapter 4: 53
Chapter 5: 105
Chapter 6: 129
Chapter 7: 147
Chapter 8: 165
Chapter 9: 187
Chapter 10: 207
Chapter 11: 219
Chapter 12: 243
Chapter 13: 257
Appendix A: 265
Appendix B: 277
Appendix C: 287
Appendix D: Desktop Power Applet Code 291
Appendix E: D-Bus: An Overview 297

Index 307

About the Authors

Ian Robert Lawrence is a Scrum Master at the Instituto Nokia de Tecnologia. He is a founding member of both the Ubuntu Brazil and Debian Amazonas communities and he is studying for an MBA in The Strategic Management of Technology Innovation at UNICAMP.

Rodrigo Cesar Lopes Belem is a free software developer and advocate who has contributed to many open source projects such as Enlightenment and Ubuntu. He has been working with free software since 2001 and currently works as a software developer at the Instituto Nokia de Tecnologia. He is studying for a Computer Science degree at the Federal University of Amazonas and holds an LPIC Level 2 certificate.

About the Contributors

Brian DeLacey worked for more than 15 years with all types of computers and software languages at companies including Lotus Development and IBM. At Harvard Business School, he spent eight years researching and writing about innovation, startups, and information technology. His interests include open source development, emerging mobile web solutions, and the future of operating systems. He holds an MBA and an A.B. in Mathematics.

Felipe Balbi has been developing Linux Kernel drivers for the last three years. He currently works for the Nokia Corporation developing the kernel for Maemo Devices and he is also an active member of linux-usb community and the USB OTG Working Group.

David Cohen is a BSc Computer Science graduate who is currently finishing an MSc degree in the same area. He's been working with open source for 8 years and has been a Linux kernel developer for the last 5. At the moment, he's working for Nokia, contributing to the development of the kernel for Maemo Devices.

Credits

Executive Editor
Carol Long

Development Editor
Kenyon Brown

Production Editor
Daniel Scribner

Copy Editor
Nancy Rapoport

Editorial Manager
Mary Beth Wakefield

Marketing Manager
David Mayhew

Production Manager
Tim Tate

Vice President and Executive Group Publisher
Richard Swadley

Vice President and Executive Publisher
Barry Pruett

Associate Publisher
Jim Minatel

Project Coordinator, Cover
Lynsey Stanford

Proofreader
Nancy Carrasco

Indexer
Ron Strauss

Cover Image
© Flying Colours Ltd./Digital Vision/Jupiterimages

Acknowledgments

Book acknowledgments are not just a list of debts the authors have racked up — many readers tend to see such a roll call of names as a code shortcut to a book's authority. As such, this book then is dedicated simultaneously both to upstream and to the Debian and Ubuntu communities — the developers, documentation writers, artists, business developers, Loco teams, and more who made this book possible.

Acknowledging my debts to the following people is not necessary, but doing so feels great: My family, especially my parents, Carolyn and Gez, who sacrificed so much to get me into a position to be able to write these words. I live to make you proud and I am profoundly grateful for everything. My beautiful fiancé, Jozi, who makes me laugh and smile, and my friends, who give their own opinions and make me think. I love you all.

Acknowledging my debts to the following people, however, is necessary. I am grateful for the complete professionalism and understanding displayed by Carol Long, Kenyon Brown and the rest of the team at Wiley — I really enjoyed working with you. This book would certainly not have happened without Brian Delacey, David Cohen and Felipe Balbi. I met Brian at UDS Boston (thanks Mark!), and he is responsible for the Mobile Linux and Mobile Directions chapters. Both David and Felipe are old work colleagues and low-level hackers who wrote the Kernel Fine Tuning and GIT sections. Superstars one and all. At the Instituto Nokia de Technology, I especially want to thank Ragner, Bruno and Milton for help with testing, and Thiago, for helping out with the review.

This section would not be complete without mentioning my co-author Rodrigo Belem. He is a gifted programmer and "muito mal elemento," too — thanks for the insights, for the many hours spent hacking, and for the friendship. Finally, this book is dedicated to the people of Brazil who love, laugh, suffer, and cry with equal passion, and to the people of Amazonas, for whom mobility is a way of life. Finally, we are slowing down and listening to what they have to say.

— *Ian Lawrence*

My participation in this book would not have been possible without the love and support of my mother and my sisters. My father gave to me love, encouragement, guidance, and the first push to learn how to fix problems. If not for him, I probably would not have followed a technology career.

Thanks to my Mom and Dad's families for all their love.

I surpassed many problems during the process of writing this book, and fundamental in this respect were (and are) Adriana Almeida and Heider Cesar. Thanks also to my fiancée for her love, attention, and encouragement in the final stages of writing this book.

In walking my path, many people appeared and were essential in influencing my hesitant steps to arrive at where I am today. These people are Guelber Menezes, Hau Wang, Bruno Monteiro, Thiago Ibiapina, and Frank Choit.

Acknowledgments

Without Ian Lawrence's invitation, maybe I would have passed my whole life without completing a work like this. Without his ideas and his willpower, maybe my life would not have taken this course. I owe you a lot, Ian. I also thank Ian for his patience and support.

At INdT I would like to thank my friends and colleagues Edisson Braga, Tomaz Noleto, and Alvaro Silva for doing my work whilst I was working on this book, and also Thiago Santos for helping us review chapters 2 and 7.

Finally, a big thanks to all the open source and free software communities, especially the Ubuntu Community.

Last but certainly not least - Granduncles Antonio and Andresson Medeiros de Melo, we will miss you.

— *Rodrigo Cesar Lopes Belem*

Contents

Introduction	**xxv**
Chapter 1: Mobile Linux	**1**
Going Mobile	1
A Short History: From Big Iron to Mighty Mouse	2
Changing Focus	3
Turning Points	4
The Generational Divide	5
Netbooks, Linux, and Ubuntu	6
A Giant's Strength in a Dwarf's Arm	8
Summary	9
Chapter 2: The Development Environment	**11**
Getting Started	12
Getting Familiar with the Ubuntu Mobile Environment	12
VirtualBox	12
KVM/QEMU	21
Using QEMU	21
ARM on QEMU	22
Using KVM	23
NETWORKING	**25**
Networking in VirtualBox	25
Networking in KVM/QEMU	26
Advanced Networking on VirtualBox and KVM/QEMU	26
Using the Bridge in VirtualBox	28
Using the Bridge in KVM	29
Sharing Files Between Guests and Host	29
Sharing Files Between Guests and Host with Advanced Networking	30
Building Your Own Virtual Image	31
Working with Images	31
Building Your Own Image	32
Installing Applications inside the Image	33
Increasing a Downloaded Image Size	34
Summary	**34**

Contents

Chapter 3: Power Management — 35

Introduction — 35
Power Saving States — 36
Power Management Packages — 36
   pm-utils — 37
      pm-suspend — 37
      pm-hibernate — 37
      pm-suspend-hybrid — 37
   pmi action — 38
   How pm-utils Works — 38
   Gnome-Power-Manager — 40
   Gnome-Power-Statistics — 40
Device Kit Power — 41
   The Quality of Service: QoS Interface — 43
Controlling Radio Transmitters — 44
   RFKILL — 45
   Bluetooth — 45
Investigating Power Usage — 46
Battery Testing — 47
   Preparing to Run the Tests — 47
   Phoronix Test Suite — 47
      Battery Comparisons — 48
      Comparing Like-to-Like — 50
Summary — 51

Chapter 4: Application Development — 53

Ubuntu Mobile Releases — 54
Creating a New Application — 55
   Application Design — 55
   Free Desktop Standards — 56
      The Desktop Entry Specification — 56
      The Desktop Application Autostart Specification — 57
      XDG Base Directory Specification — 57
      Desktop Menu Specification — 57
   Hildon: An Application Framework for Handheld Devices — 58
   What Is Hildon in Terms of Code? — 58
      Creating the Program — 58
      Menus — 59
      Toolbars — 59
      Window-Specific Settings — 59
      Program-Wide Settings — 59

Contents

Hibernation	60
Putting Hildon Together	60
Hello World	60
Other Toolkits	64
Signals	65
Layout	66
Horizontal Boxes	66
Vertical Boxes	66
Glade	66
Handling the .glade File	69
Clutter	70
QT	75
EFL	79
Canola	81
Elementary	81
What Key Technologies Do I Need to Know to Develop Applications for a Mobile Device?	**82**
D-Bus	82
Object Paths and Bus Names	83
Exporting Objects with D-Bus	84
Connect to a D-Bus Signal	85
Useful D-Bus Command-Line Applications	85
D-Bus Viewer	85
D-Bus Send	87
D-Bus Monitor	87
D-Bus Launch	88
D-Feet	88
D-Bus Security	89
PolicyKit	90
GConf	**91**
Notifications	93
Putting All the Concepts Together	**94**
Summary	**103**

Chapter 5: Application Packaging — 105

Background and Important Tools	**105**
Packaging and Using a PPA	**108**
Initial Debianization	**109**
rules	109
changelog	112
control	112

xvii

Contents

copyright	113
Other Debian Files	114
Building the Package	115
Uploading to a PPA	115
REVU	116
RFA Packages	117
Creating Your Own Repository	**118**
Simple Repository	118
Automatic Repository	118
Setting Up a Repository	118
Adding Packages to a Repository	119
Removing Packages From a Repository	119
Backporting KVM	119
PBuilder	**120**
Configuring PBuilder	120
Performing Actions on PBuilder	123
Creating a Distribution Environment	124
Building a Package to a Specific Release	124
Updating the PBuilder Environment	124
Using pdebuild	124
Configuring Actions	125
Additional Hook Manipulation with PBuilder	125
Hook Script Resource	126
Mount Bind a Package Repository for Use with PBuilder	126
Ubuntu Policy	126
Categories	126
Sections	127
Summary	**127**

Chapter 6: Application Selection   129

Business Users	**129**
Documents	130
A Practical Example	131
Multimedia Users	**131**
A Practical Example	132
Useful Keybindings in the Entertainer GUI	133
Social Network Users	**134**
A Practical Example	134
Set Up the Environment	136
Copy the Gadget	136
Modify It	136

Contents

Location-Aware Users — 138
  A Practical Example — 138
    Background — 138
    Implementation — 139
    Test the Gypsy to GPS Connection — 142
    Interaction with the GPS Daemon — 142
    D-Bus and HTTP Requests — 143
Summary — 146

Chapter 7: Theming — 147

What Is a Theme? — 147
  Where Are Themes Located in the Filesystem? — 148
  What Is a Theme Engine and Where Are They Located? — 148
  Theming Ubuntu MID — 148
    What Happens When a MID Device Boots? — 149
Modifying Themes — 150
  A Useful Tool When Working with Themes — 150
  Theme Structure — 151
    The theme.xml File — 151
    The gtkrc File — 152
    Customizing a gtkrc File — 152
    Padding — 153
    Styles — 153
    Colors — 153
    Applying the Style — 154
  Theming Ubuntu MID — 154
    Manually Theming MID — 154
    Automatically Theming MID — 156
  Theming Ubuntu Netbook Remix — 157
    Boot Splash — 158
    Creating a gdm Theme — 158
    Customizing the Netbook Launcher — 159
Performance Testing of Themes — 160
  Test the Human Metacity Theme — 160
    Comparisons — 160
    X Window Testing — 162
Summary — 163

Chapter 8: Kernel Fine-Tuning — 165

Ubuntu MID Kernel Overview — 165
Kernel-Tuning Methods — 165
  Create an Ubuntu Package — 166
  Create a Debian Package — 172

Contents

Updating a Customized Kernel Tree	**175**
Updating an Ubuntu Kernel Tree	175
Update a Non-Ubuntu Kernel Tree	181
Dynamic Kernel Module Support	**181**
Inside the DKMS Framework	182
Basic DKMS Commands	183
Summary	**186**

Chapter 9: Testing and Usability — 187

Why Test?	**187**
Ubuntu Desktop QA	**188**
Mago — A Desktop Testing Initiative	188
Building an Application for Testing	**189**
Getting Started	189
Application Creation	190
Testing with Mago	193
Adding the Browser Test to Mago	193
Linux Standards Base and Certification	196
Installing the LSB Application Testkit	196
Running the LSB Application Testkit	196
Other Testing Tools	**197**
Phoronix Test Suite	197
PBuilder for Automating the Testing of Packages	200
Other Useful Linux Performance Testing Tools	201
ps	201
top	201
time	201
procinfo	201
free	202
memstat	202
memcheck and Valgrind	202
Latencytop	203
Testing Strategies	**203**
Basic	203
Advanced	204
Compliance	204
Bug Reporting	**205**
If You Find a Bug . . .	205
Filing a Bug Report Automatically	206
Reporting a Bug from the Command Line	206
Summary	**206**

Contents

Chapter 10: Tips and Tricks — 207

Improving Boot Speed — 207
Hard Coding Modules — 207
Creating a /tmp That Is Half the Size of Physical RAM — 208
Energy Tips — 208
   Recharging Correctly — 209
   Laptop Mode — 209
   Getting to Know the Battery on a Device — 210
   CPUFREQ and Governors — 211
   Use Power Management Settings on Disks — 211
   Disabling atime — 212
Turning Off Background Services — 212
Adobe Flash — 213
Configuring the Touchscreen — 214
Watching Hard Disk Activity — 217
Summary — 218

Chapter 11: Putting It All Together — 219

Important Things to Consider — 219
   Check If the Device Architecture Is Supported by Ubuntu — 220
   Checking the Hardware — 220
   Fine-Tuning the Kernel — 221
   Defining Power Policies — 221
   Is It an Embedded System? — 221
Customizing the User Interface — 222
   Boot Selector — 222
   Display Manager — 223
      GDM — 224
      Pre-Configuring GDM — 224
   Setting the Default Ubuntu, XFCE, and Hildon Behaviors — 224
Fine-Tuning the Build Process — 225
   Setting Up a Repository — 225
   Caching Packages with approx — 225
Creating a Default Ubuntu Image — 226
   Choosing Which Type of Installer to Use — 226
      When to Use Debian-Installer (Ubuntu Alternate Image) — 226
      When to Use Ubiquity (Ubuntu Desktop Image) — 226
   Getting Started on the Image: Preparing the Environment — 226
      Finally, Building the Default ISO — 229

Contents

Building a Customized Ubuntu Image — **230**
Inside Seed Germination — 231
Germinating the Seeds — 232
An Example: Germinating Ubuntu Netbook Remix — 234
Packages and Repositories — **235**
Generating Metapackages the Ubuntu Way — 235
Building the metapackage — 236
Generating Metapackages the Simple Way — 238
Preseeding the Installer — 239
Adding Packages to the Image — 240
Finally, Build the Custom ISO — 240
Ubuntu Policies, Trademarks, Copyright, and Common Sense — **240**
So What Is a Derived Distribution? — 241
When to Use the LGPL — 241
Summary — **242**

Chapter 12: Mobile Directions — **243**

Choice, Change, and Opportunity — **244**
Evolution and Software Development — **246**
Darwin — 247
Mendel — 247
Mayr — 248
Frankenstein — 248
Big Ideas to Think About — **249**
The Politics of Technology — 249
The Next Billion — 249
Sensory Overload — 249
Cloud Computing — 250
ARM Wrestling — 250
Razors and Blades — 251
Free Lunch — 251
Computing on the Edge — 251
The Future — **253**
Ubuntu, Linux, and Mobile Computing — **254**
Summary — **255**

Chapter 13: Common Problems and Possible Solutions — **257**

The Boot Process Stops — **257**
Application Icon Does Not Appear — **258**
Performing a Dual Boot — **259**
Setting a Flag Automatically — **259**

Using USB	260
Running Ubuntu on Freerunner	260
Running Ubuntu on Arima	261
Ubuntu Intrepid UMPC Project	261
Installing Ubuntu Netbook Remix on a UMPC	261
Using apt	261
Joining the Ubuntu Mobile Developers Team	262
Using KVM or QEMU	262
Graphical Corruption	262
Poor Performance	263
Summary	263

Appendix A: Ubuntu's Right ARM — 265

Appendix B: Git Usage — 277

Appendix C: Hosting Your Project on Launchpad — 287

Appendix D: Desktop Power Applet Code — 291

Appendix E: D-Bus: An Overview — 297

Index — 307

Introduction

Professional Ubuntu Mobile Development is designed for all developers interested in a practical, hands-on way of learning development on mobile devices. The book is designed to show you how to complete real-world tasks in efficient and often, we hope, innovative ways. Our goal is that the examples in this book will help you understand the techniques you need when working with Ubuntu Mobile. Our hope is that you can then creatively apply them to your own real-world problems.

As such, the book is not a "static" object but we as authors have tried to model this dynamism through an emphasis on discrete, reusable units of logic — the chapters — the names of which became obvious as we worked at customizing Ubuntu Mobile. This means the book can be approached in numerous ways.

Whom This Book Is For

The book is for developers with some experience working with Debian-like systems such as Ubuntu, but it is also for developers with experience with other operating systems, who perhaps want to explore or want to rapidly come up-to-speed with the key platform features of Ubuntu Mobile. It is well suited for developers who are "perfectionists with deadlines" and, as such, is not an introduction to either embedded development or Ubuntu. To get the most from the book, you should understand programming principles and have a healthy dose of curiosity about how things work and how to adapt the examples provided to your particular situation and demands.

It is also not a book about how to install Ubuntu onto any particular mobile device (there are other guides available on the Internet for this), but it is a book which will be useful if you would like to use Ubuntu Mobile in its various flavors for your own customization projects.

What This Book Covers

On August 28, 1909, *The North-China Herald* published an article called "The Moving Target Problem," which concerned the growing popularity of moving targets at the expense of the bull's-eye when training riflemen. This was much to the chagrin of the military, which maintained that the best rifleman in actual warfare would be the one who had had careful training on the bull's-eye and had from his earliest career sought to observe and then rectify his errors in marksmanship.

Anyone who has written a book will certainly side with the military on this one. The scope of Ubuntu is large and the mobile project so new that some things in this book might well have changed when you read them. What this book gives is a snapshot of current best practices and the tools you need at hand to implement them.

The most important thing when developing for an embedded device is to have a development environment set up. Once this is done, it is possible to develop, package, and test your application in an environment that provides a reasonable approximation of a real device. This is then followed by

Introduction

several chapters, each of which emerged from a real-world situation during the course of our work with mobile device development and "customixation." The chapters are the kinds of things we wanted to see when deadlines were looming and we hope that together they make up the kind of book we wish we had on hand when we were starting out.

We immensely enjoyed writing this book and hope that you will enjoy reading it just as much.

How This Book Is Structured

The book covers the following topics:

Chapter 1, Mobile Linux — discusses the possibilities and probabilities of running Linux on a billion devices.

Chapter 2, The Development Environment — steps you through setting up a work environment.

Chapter 3, Power Management — examines the greater divergence in the future between performance-optimized and power-optimized mobile devices.

Chapter 4, Application Development — discusses developing applications for mobile devices.

Chapter 5, Application Packaging — illustrates preparing your applications for distribution.

Chapter 6, Application Selection — is about choosing the right applications for your target users.

Chapter 7, Theming — shows you how to customize the look and feel of a mobile device.

Chapter 8, Kernel Fine-Tuning — explains how to represent the band of software nearest to the hardware.

Chapter 9, Testing and Usability — discusses meeting benchmarks and standards for stability and performance.

Chapter 10, Tips and Tricks — suggests ways to save time with some hard won advice.

Chapter 11, Putting It All Together — discusses the benefits and potential pitfalls of having your own mobile distribution, and shows how to actually seed and germinate the image.

Chapter 12, Mobile Directions — discusses the future of Linux on mobile devices.

Chapter 13, Common Problems and Possible Solutions — identifies problems and provides solutions.

What You Need to Use This Book

The book was written and the code was developed on computers running Ubuntu Jaunty and Ubuntu Karmic.

The code in the chapters was tested and used variously on an Acer Aspire One, Geode, Eee PC, and a Beagleboard. An actual device or board like these is obviously nice but not essential to fully enjoy this book.

History and Background to the Ubuntu Mobile Project

This background comprises various conversations on IRC at #ubuntu-mobile, and it now forms the basis of the wiki page at `https://wiki.ubuntu.com/MobileTeam/Mobile/History`.

The project started with an announcement at the start of the Ubuntu Hardy release cycle by Ubuntu CTO Matt Zimmerman to the ubuntu-devel mailing list, which said:

> *It is clear that new types of device — small, handheld, graphical tablets which are Internet-enabled — are going to change the way we communicate and collaborate. These devices place new demands on open source software and require innovative graphical interfaces, improved power management and better responsiveness.*
>
> *To fulfill the aims of our mission and in response to the technical challenges that these devices pose, we are announcing the Ubuntu Mobile and Embedded project.*

The Ubuntu Mobile and Embedded Project

This was an extension of the GNOME Mobile and Embedded Initiative (GMAE). The naming of the Ubuntu Mobile and Embedded Project caused some confusion at the time mainly because there existed a previous Ubuntu Embedded Team. This team was working toward the creation of tools, documentation, and a binary/source release for the purpose of running Ubuntu on small PC style hardware and true embedded devices. The Ubuntu Embedded team was mainly looking at getting Ubuntu working on ARM processor–based devices.

The Ubuntu Mobile and Embedded Team then started work during the Ubuntu Hardy release cycle and as time passed, most of the focus during this cycle was on getting *Hildon* and *Moblin* into Ubuntu, and porting to the new *LPIA* architecture from Intel:

- **Hildon** is an application framework for mobile devices developed by Nokia and now a part of GMAE, which focuses on providing a finger-friendly interface. It is primarily a set of extensions that provide mobile-device–oriented functionality.

- **Moblin** is an Intel-sponsored, open-source community and application framework to create consumer-friendly applications and user interfaces across a range of Mobile Internet Devices (MIDs), netbooks, and embedded devices.

- **LPIA** (Low Power Intel Architecture) is a new processor architecture, which, while resembling i386, uses different optimization options in the compiler and different configuration and build options for some packages. Specifically, LPIA uses GCC-4.2 as the system compiler (instead of GCC-4.1 which is used for the other Ubuntu architectures).

Introduction

The Ubuntu Hardy Release

During the Hardy cycle, then, there was work toward getting Ubuntu working on various devices, based on both the *Menlow* and *McCaslin* platforms from Intel:

- **McCaslin** preceded Menlow. It contains an Intel A100/A110 processor. An example of a device using this processor is the Samsung Q1 Ultra.

- **Menlow** is the name of the platform that contains an Intel Atom processor (codenamed Silverthorne and Diamondville). It uses Poulsbo Chipset (aka System Controller Hub). An example of a device using this processor is the Aigo MID. This work didn't complete within the Hardy cycle, and continued in a *PPA*.

- **PPA** (Personal Package Archive) is a way for developers to build and publish packages of their code, documentation, artwork, themes, or any other contribution to free software.

This resulted in the first release image of Ubuntu MID (which was some time after the official Ubuntu Hardy release).

While this release was targeted for 4 to 6-inch screen devices, most of the testing and the work was done on a Samsung Q1 (known as the reference device) or on somewhat awkward development kits.

The Ubuntu Mobile Team

Some time after the Hardy developer alpha release there were some nomenclature cleanups within the Ubuntu project. As a result of this, the Ubuntu Mobile and Embedded Team became the Ubuntu Mobile Team, as the goals were increasingly divergent from those of the Ubuntu Embedded Team and of "pure" embedded development itself.

Netbook Remix

Completely independently, a group of Canonical developers from the *Original Equipment Manufacturer* (OEM) team was looking at technologies to support the new "netbook" devices. *OEM* refers to companies that make products for others to repackage and sell.

Using a mix of GNOME, OpenHand code, and some of the work that went into Ubuntu MID, and additional development, they produced the "Netbook Remix," based loosely on Ubuntu Hardy.

The resulting Netbook Remix was released before the first release of Ubuntu MID. More about the reasoning and ideas behind the Netbook Remix can be found at `http://www.markshuttleworth.com/archives/151`.

> The Netbook Remix is based on Ubuntu Long Term Support (LTS) releases such as Hardy.

Introduction

UDS Intrepid

At the UDS (Ubuntu Developer Summit) for Intrepid, there were demonstrations of preliminary versions of both the Netbook Remix and Ubuntu MID. There were also demonstrations of additional work based on the Ubuntu MID done by the Canonical OEM team. Along with this were further demonstrations done of the Edubuntu CMPC image. It was decided that some consolidation was necessary.

As a result of all of this, the Ubuntu developers decided that there would be specific Ubuntu Mobile releases for Intrepid.

Intrepid

So, for Intrepid, there was work done to create Ubuntu Mobile. The Ubuntu Mobile release was designed for larger 7–9-inch screens, which work best on the Samsung Q1 (as that happened to be the 7–9-inch hardware owned by the developers working on this release flavor).

The Ubuntu Mobile releases do not have any of the hardware settings or hard-coded configurations that were present in the Hardy-based Ubuntu MID, and so should not be nearly so tied to the specific device. Indeed, one of the specific goals for Intrepid for Ubuntu MID particularly was to move away from the hard coding, and so enable a wider variety of devices (perhaps including appropriately sized hardware such as the Aigo MID or Sharp D4).

Intrepid has an ubuntu-mid.img as well as an ubuntu-mobile.img. Images are officially built in the Ubuntu infrastructure on cdimage.ubuntu.com, which results in daily ubuntu-mobile images. Although the ubuntu-mobile release was a beta one, it was well reviewed on UMPCPortal (www.umpcportal.com/) and generated some community interest.

Jaunty

The Jaunty Ubuntu Mobile release was based on Hildon 2.2, which was itself based on Clutter and GTK. This brought significant improvements for Ubuntu Mobile. These include enhanced GTK+ widgets to make finger-friendly interfaces, while staying compatible with API calls in existing code and an optional/complementary user interface library called Clutter. Clutter, discussed in Chapter 4, is an OpenGL ES rendering library for creating visually rich and interactive user interfaces.

Karmic

The karmic release for Ubuntu Mobile is focused on the Ubuntu Netbook Remix and ARM releases of Ubuntu (see below). In addition, a one-off Ubuntu Moblin Remix was rolled out primarily for demonstration and comparison purposes, but also to showcase the best of both projects.

Also during this cycle, Ubuntu MID became community-maintained in the Ubuntu Liquid Remix project. If you would like to join this project, go to https://edge.launchpad.net/~ulr. The first community release will be based on Ubuntu Lucid, which will be an LTS version from Ubuntu.

Ubuntu ARM

On November 13, 2008, Canonical, the commercial sponsor of Ubuntu, announced that in response to demands from device manufacturers the Ubuntu operating system would be available on the ARMv7 processor architecture. This includes both the ARM Cortex-A8 and Cortex-A9 processor-based systems, which are the highest performing, most power-efficient processors released to date from ARM.

Introduction

ARM is used in a wide range of devices such as the Nokia N700/N8xx series, the Sharp Zaurus, the Linksys NSLU2 Network Attached Storage (NAS) device, and the iPhone, which uses an older ARM v6 chip. Ubuntu ARM became officially available in April 2009.

Conventions

Instead of including a lot of numbered steps, instructions are provided with the code displayed in the text, as described in the example that follows.

To use QEMU, install the necessary packages:

```
$ sudo apt-get install qemu kqemu-common kqemu-source
```

Next, add your user name to the kqemu group. Adding your user name to the kqemu group means that you do not need to run QEMU as the root user.

kqemu is the name of the module that is used to accelerate QEMU.

To add your user name, run

```
$ sudo adduser <username> kqemu
```

To make these changes effective, it is necessary to log out and then log back in again.

To test QEMU, go to the folder that contains the image that you downloaded from http://cdimage.ubuntu.com and run

```
$ qemu -localtime -m 384 -boot d ubuntu-9.04-mid-lpia.img
```

Source Code

As you work through the examples in this book, you may choose either to type in all the code manually or to use the source code files that accompany the book. All of the source code used in this book is available for download at http://www.wrox.com. Once at the site, simply locate the book's title (either by using the Search box or by using one of the title lists) and click the Download Code link on the book's detail page to obtain all the source code for the book.

Because many books have similar titles, you may find it easiest to search by ISBN; this book's ISBN is 978-0-470-43676-9.

Once you download the code, just decompress it with your favorite decompression tool. Alternately, you can go to the main Wrox code download page at http://www.wrox.com/dynamic/books/download.aspx to see the code available for this book and all other Wrox books.

Introduction

Errata

We make every effort to ensure that there are no errors in the text or in the code. However, no one is perfect, and mistakes do occur. If you find an error in one of our books, such as a spelling mistake or faulty piece of code, we would be very grateful for your feedback. By sending in errata, you may save another reader hours of frustration and at the same time you will be helping us provide even higher quality information.

To find the errata page for this book, go to `http://www.wrox.com` and locate the title using the Search box or one of the title lists. Then, on the book details page, click the Book Errata link. On this page, you can view all errata that has been submitted for this book and posted by Wrox editors. A complete book list including links to each book's errata is also available at `www.wrox.com/misc-pages/booklist.shtml`.

If you don't spot "your" error on the Book Errata page, go to `www.wrox.com/contact/techsupport.shtml` and complete the form there to send us the error you have found. We'll check the information and, if appropriate, post a message to the book's errata page and fix the problem in subsequent editions of the book.

p2p.wrox.com

For author and peer discussion, join the P2P forums at `p2p.wrox.com`. The forums are a Web-based system for you to post messages relating to Wrox books and related technologies and interact with other readers and technology users. The forums offer a subscription feature to e-mail you topics of interest of your choosing when new posts are made to the forums. Wrox authors, editors, other industry experts, and your fellow readers are present on these forums.

At `http://p2p.wrox.com` you will find a number of different forums that will help you not only as you read this book, but also as you develop your own applications. To join the forums, just follow these steps:

1. Go to `p2p.wrox.com` and click the Register link.
2. Read the terms of use and click Agree.
3. Complete the required information to join as well as any optional information you wish to provide and click Submit.
4. You will receive an e-mail with information describing how to verify your account and complete the joining process.

> You can read messages in the forums without joining P2P, but in order to post your own messages, you must join.

Once you join, you can post new messages and respond to messages other users post. You can read messages at any time on the Web. If you would like to have new messages from a particular forum e-mailed to you, click the Subscribe to this Forum icon by the forum name in the forum listing.

For more information about how to use the Wrox P2P, be sure to read the P2P FAQs for answers to questions about how the forum software works as well as many common questions specific to P2P and Wrox books. To read the FAQs, click the FAQ link on any P2P page.

Professional
Ubuntu® Mobile Development

Mobile Linux

This chapter introduces mobile computing in the context of the evolution of different computer types. More important, it presents reasons why developing mobile applications with Linux and Ubuntu makes economic and technical good sense.

> *More than three-quarters of the expert respondents (77%) agreed ... that the mobile computing device — with more significant computing power in 2020 — will be the primary Internet communications platform for a majority of people across the world.*
>
> Pew Internet & American Life Project, The Future of the Internet III, December 14, 2008

Going Mobile

Since the first computers were created, there has been a constant push for smaller, faster, cheaper systems that provide more personal power. In December 2008, quarterly laptop sales outnumbered desktop computer sales for the first time ever. Netbook computers — smaller than laptops, with a price performance profile that took the market by storm — were the unexpected hit of 2008. Consider the following statistics: International Data Corp (IDC) estimates 20.6 million netbooks will ship in 2009 (compared to 137 million full-sized laptops). ABI Research says that number could reach 35 million in 2009 and 139 million in 2013. Ultra Mobile PCs (UMPCs) seem to be trickling along at one or two million. Mobile Internet Devices (MIDs) are projected to see a healthy jump in sales, with some estimates placing sales at nearly 6 million in 2009 and triple that in 2010 — not yet the runaway success of netbooks, but still substantial in comparison to smartphones. As demand for mobile solutions has grown, Linux and Ubuntu have improved. Today's mobile markets — for both end-users, and vendors who look to bundle an operating system with their hardware — align well with Linux and Ubuntu.

Let's take a quick look at how these markets and technologies evolved, and why Linux and Ubuntu are primed to deliver mobile solutions.

Chapter 1: Mobile Linux

A Short History: From Big Iron to Mighty Mouse

The Harvard Mark I computer that's shown in Figure 1-1 was 51 feet long and 8 feet high. It first booted up in 1944 in order to multiply, divide, do logarithms, and process trigonometric functions. This system was widely viewed as the beginning of the modern computer era. Imagine, a computer that could do only five multiplication problems and two division problems a minute! Logarithmic processing was a good time to go out for a coffee break.

Figure 1-1

Since that time, there has been an incredible evolution of computing technology following the steady path of Moore's Law, which is shown in Figure 1-2.

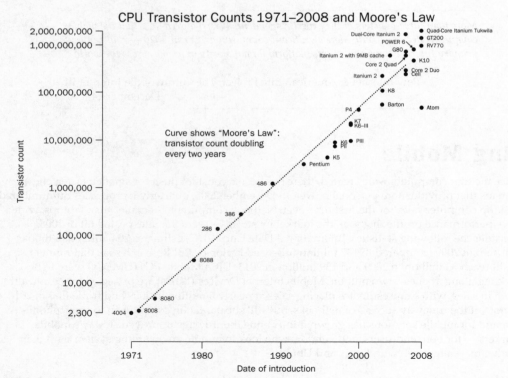

Figure 1-2

Chapter 1: Mobile Linux

The major eras in computer technology can be classified as follows:

1960s — Mainframes ("Big Iron") and minicomputers (multiuser, interactive)

1970s — Personal/desktop computer: microprocessor-driven, installable applications

1980s — Luggables: moveable computers, weighing 15–30 pounds

1990s — Laptops followed by slimmed down notebooks: 14-inch screen or larger

2000s — Subnotebooks: 12–13-inch screen, Portable Media Players (for example, iPod)

2007–2010 — Netbooks: 7–10.2-inch screen; Mobile Internet Devices (MIDs): 4–7-inch screen

The year 2000 marked the beginning of the cell phone and smart device era. The Nokia 9210 Communicator and RIM Blackberry hit the market with compact offerings that were phones but also much more. They weren't general purpose computers, but you might have gotten the sense that that's what they wanted to be when they grew up.

Changing Focus

Two of the defining characteristics in the evolution of computers have been physical and display size. These defining attributes influence user interactions as well as developer strategies and solutions.

Apple opened the door on the smallest consumer-oriented computing segment with its insanely successful introduction of iPods in October 2001. The first iPod was a computer stick the size of a small pack of gum. Consumer electronics have a tendency to get smarter through the magic of software and soon the iPod line represented complete media devices.

In January 2007, Apple completely blew the roof off this segment by introducing the iPhone. Later in the year, the iPod Touch was introduced. These portable electronic devices had powerful operating systems under the hood — variants of OS X specially made for the hardware. Elegantly packaged, with WiFi, and high-resolution touchscreens, they now ran applications like those on notebook computers. Apple was effectively shipping what would come to be called a Mobile Internet Device (MID).

Loyal Apple customers stood in long lines to snap up the first shipped products. International customers paid huge premiums for early shipments. Enthusiastic developers created thousands of new applications . . . even though it initially seemed as though Apple didn't want them to. An ecosystem of telephone carriers, music providers, and accessory makers helped grow the market.

Apple sold an astounding number of iPods — more than 173 million of these very mobile gadgets by September 2008. You've seen them everywhere — on trains, planes, and buses, and in gyms, schools, office buildings, and sports venues. (Goodness, I have one friend who owns nine of these things!)

When Apple introduced the Macintosh in 1984, they wrote the bible of good human interface design. In releasing the iPhone and iPod Touch, Apple rewrote the book on user interface design for small mobile computers. The iPhone redefined what was possible for small computer packaging. Consider the specs: 3.5-inch diagonal multi-touch display (480×320 pixel resolution at 163 ppi), less than 5 ounces, in slim packaging.

Chapter 1: Mobile Linux

It didn't take long for thousands of developers to build creative, high-quality applications for the iPhone and iPod Touch platform. The 2.0 generation of Apple's software (and SDK) generated tremendous developer excitement. In March 2008, Apple announced a beta program for the iPhone 2.0 SDK — the ease of use of the development platform, and the impressive application results drove 25,000 developers to try and get in on the beta. Apple managed to share all their SDK APIs widely across product lines — the iPhone SDK has great commonality across the OS X kernel. On June 9, 2008, Apple announced a record number of 5,200 developers at their Worldwide Developer Conference — the first sellout in 25 years of the event.

Customers bought and installed software by the millions. In the first six months of operation (from July 11, 2008 to January 18, 2009) the App Store saw 500 million downloads from its catalog of 15,000 applications. Apple created a mobile user interface that broke new ground, offering numerous examples and lessons to Linux and Ubuntu Mobile developers.

Turning Points

The financial crisis that began in September 2008 tripped up economies around the world. Big software companies, leading hardware manufacturers, and dominant component manufacturers were all affected. At the beginning of 2009, the mobile phone business began to slow down. *TechCrunch* reported that the "top five cell phone manufacturers (Nokia, Samsung, LG, Sony Ericsson, and Motorola) dropped 13 percent year-over-year in the fourth quarter of 2008. Unit shipments decelerated from 14 percent growth in the second quarter to 2 percent growth in the third quarter, and then finally went into negative territory in the fourth quarter." The article's author asked, "Are cell phones no longer a growth business?"

Even Apple's idyllic iPhone fell off the selling cliff. Apple sold 6.9 million units in the September 2008 quarter, but that fell by more than 25 percent to 4.4 million in the December quarter. At the same time, RIM made a big-splash introduction — backed by a $100 million marketing campaign — of its highly anticipated BlackBerry Storm. It was judged by many observers to be a relative flop compared to the iPhone introduction: Blackberry sold a half-million units in the first month.

Cell phones and laptops dominated tech talk in the first five years of the twenty-first century, but it could be a very different picture over the course of the next decade. The same day those declines in cell phone sales were being reported, a lead story in *BusinessWeek* rallied excitement around a promising area of growth: "Intel Readies Push into Mobile Internet Devices."

While phones were getting put on hold, notebook (and mobile computer) sales were rising. The December 2008 *Wall Street Journal* reported quarterly sales of notebook computers exceeded that of desktop sales for the first time ever:

> *World-wide shipments of notebook computers rose nearly 40% from last year to 38.6 million units as desktop shipments fell 1.3% to 38.5 million units.*

Another bright light in the numbers was the brand new category of netbook computers:

> In the first months of 2008, netbooks were less than two percent of laptop sales. By December, that had shot up to 12 percent of total unit volume — and accounted for almost two-thirds of the sales increase in the entire laptop category.

The *Financial Times* put some numbers to that early in 2009 when they sized up the market:

> Netbook sales have grown from about 350,000 units in 2007, when Asustek introduced its first models, to 10m in 2008. The CEA predicts unit sales will rise 80 percent to 18m in 2009, in spite of the global downturn.

Ubuntu and Linux are right in the middle of a burgeoning mobile marketplace. And there are numerous developer opportunities ahead.

The Generational Divide

If you have any doubt about the promise of Ubuntu as an operating system, or the likelihood that it will gain a significant market presence in the future, you may find this short story of interest.

Sometimes people evaluate a computer's ease of use by asking: "Is it easy enough for your grandmother to use?" A better question is this: "Is it easy enough for a 15-year-old to use?" I love my grandmother dearly, but even if she did have an interest in computers, I doubt big business would be wise in building a five-year business plan around her particular usage scenarios. On the other hand, a 15-year-old typically has the interest and the need, and presents an attractive lifetime value proposition.

My 15-year-old son came to me one evening carrying his Windows-powered notebook. "Hey, Dad, can you take a look at this and help me get WiFi working?" I anticipated what had happened with his sister's notebook — it was time for another Windows reinstallation.

He turned the laptop around to show me the screen and what I saw was shocking: the Ubuntu desktop.

"I installed Ubuntu," he said. "Wow, the browser is super fast when I plug in the Ethernet. Now I remember what the Internet was like." He'd been fighting the gradually degrading performance of his Windows Vista installation for weeks now and finally just decided to fix it.

His notebook was just a year old — with plenty of horsepower and a price tag that pushed it over $1,000. It had all the latest technologies — WiFi card, hard disk, a nice screen. But it had slowed noticeably and inexplicably. My gut reaction was to reinstall Windows for him. I made the offer despite the dread I felt anticipating a multi-day effort to get Windows working again. Chasing down drivers for Windows is no fun.

"What happened?" I asked. "I can reinstall Windows for you."

Chapter 1: Mobile Linux

"No need for Windows," he said. "Ubuntu is fine. It has a word processor. I don't need any of my old files. Vista had gotten so slow I just couldn't use it. Last night I was trying to use Dictionary.com and it never worked. I had to go to the downstairs computer to look up a word. It would have been faster to use a printed dictionary. That was the moment I decided to install Ubuntu. I used the CD you left lying around. No problems."

I was flabbergasted. I'm not sure what shocked me more — that he had installed Ubuntu by himself without assistance, or that he had installed it without letting me know in advance.

He continued, "Vista had gotten so slow it would take like 30 minutes to start up. I had disabled all the Vista applications that run at startup, but that didn't help. I resorted to leaving the system running all the time to avoid the delay in restarting."

Just like that — he'd made the switch. No regrets. No remorse. No difficult separation. He just wanted a better, quicker, more reliable system. And he knew where to find a free alternative.

If Windows Vista were to lose the teenage marketplace, it would certainly be in *real* trouble. On the other hand, if Ubuntu starts winning the 15-year-old customer segment, it will be in really good shape.

Netbooks, Linux, and Ubuntu

The ASUS Eee PC shipped in September 2007. It was in instant hit. Small, light, and inexpensive (under $300), it ran Xandros Linux as well as Windows XP. After Intel's Atom processor was released in 2008, the Eee PC switched to Atom — an energy efficient chip with a well known, powerful instruction set.

An entirely new classification of computer had been born — the netbook. Other vendors soon followed with their own offerings: HP Mini-Note, MSI Win, Acer Aspire One, Dell Inspiron Mini 9, and the Lenovo IdeaPad S10 were offered in varying configurations. The Information Network estimates that "11.4 million netbooks were sold in 2008, up from 400,000 in 2007." For 2009, the firm estimates that netbook sales will grow 189 percent to 21.5 million. Meanwhile, the firm estimates that 145.9 million notebooks were sold in 2008 and projects that number will grow 21.8 percent in 2009 to 177.7 million. Other projections suggest a day when these small form factor computers will outsell notebooks. According to a LinuxDevices report, ABI Research predicted that "35 million netbooks will ship this year, rising to 139 million in 2013.

I watched as the Eee PC became a cult hit through December 2007 and into 2008. Eventually, I bought an Acer Aspire running the new Intel Atom chip. ("By 2010, Atom will be competitive in every aspect of mobile computing," according to Intel's Senior Vice President Pat Gelsinger.)

This little Acer included a solid state drive in a slim, durable, nice looking package. It seemed like a wonderfully mobile computer. In Figure 1-3, you can see the relative size of the Acer Aspire One compared to a T-Mobile smartphone, iPod Touch, and Compal MID — all sitting atop a Dell Inspiron notebook computer.

Chapter 1: Mobile Linux

Figure 1-3

The Aspire One came installed with a customized operating environment built on top of Linux. Since August 1991, when Linus Torvalds uploaded his first few modules for his new operating system to Usenet, Linux has been evolving into a robust, state-of-the-art operating kernel.

Most operating system distributions that are based on the Linux kernel are basically modified versions of the GNU operating system.

In addition to providing a strong software foundation for distributions, Linux is like the Swiss Army Knife of software. (I've used it several times to recover files from computers that were running Windows but had become corrupted and could no longer boot.) It's been ported to a large number of hardware platforms and increasingly, vendors are developing their hardware drivers for open source and Linux.

Thousands of Linux-based distributions are available — but I decided to install Ubuntu. The resulting system provides a full Internet and computing experience.

Why Ubuntu? Quality, reliability, and widespread adoption are a few good reasons. Ubuntu and Debian were the first two distros I began working with in 2005, and I have stuck with them.

Chapter 1: Mobile Linux

Month after month Ubuntu has been at or near the top of the DistroWatch list of Linux distributions. Moreover, there is a great developer community. The healthy (and wealthy) coordination by Canonical provides stability and important direction. Early in 2009, the *New York Times* profiled Canonical's founder and Ubuntu visionary, Mark Shuttleworth:

> *Created just over four years ago, Ubuntu (pronounced oo-BOON-too) has emerged as the fastest-growing and most celebrated version of the Linux operating system, which competes with Windows primarily through its low, low price: $0.*
>
> *More than 10 million people are estimated to run Ubuntu today, and they represent a threat to Microsoft's hegemony in developed countries and perhaps even more so in those regions catching up to the technology revolution.*
>
> *'If we're successful, we would fundamentally change the operating system market,'*
> *Mr. Shuttleworth said during a break at the gathering, the Ubuntu Developer Summit. 'Microsoft would need to adapt, and I don't think that would be unhealthy.'*

It's easy to grab and go with a netbook. And whatever I can do on a desktop PC, I can do on this device — not as fast, but for many things it works just fine. As a result, a netbook running Ubuntu becomes an entirely new platform to develop for, but it is a very familiar platform. Yet the range of application possibilities is vastly greater because of all the different dimensions added by mobility.

So why not start developing for this category of mobile computer, using Linux and Ubuntu? If you do, you'll be in good company. Consider what Linus Torvalds had to say early in 2009:

> *It's a huge job to do a distribution. The reason there are hundreds is it is easy to start your own, but if you want to be a leader and introduce new code, the testing and Q&A involved is enormous. It depends on having enough users that you get coverage and it is unreasonable to expect too many large distributions. Ubuntu grew surprisingly quickly and maybe that can happen again . . . .*
>
> *I was doing kernel development on a netbook and it was not at all horrible. The screen was too small, but we are getting to a stage where you can get a cheap good laptop.*
>
> *A few years ago you could get a small netbook but it would be twice the cost. The netbook market changed the game — they are not seen as an executive toy, but a low-end laptop which is much healthier.*
>
> *With netbooks a lot of the desktops have trouble going to smaller screens. All of a sudden you can't press the okay button because it's outside the screen. As screens go as small as phones, Google's Android could be a contender for netbooks so you may see Android growing up instead of desktops growing down.*
>
> *We are in the first phase of netbooks and there are some teething problems. The dumbed-down interface was a teething problem and the first netbooks were underpowered.*

A Giant's Strength in a Dwarf's Arm

During World War II, one of the greatest scientific innovations involved radar. The British had developed crucial underlying technologies and fundamental understanding of microwave physics, far beyond what scientists in the United States had discovered. One of the British inventions was something called a *cavity magnetron*. This device led to the successful development of a light, compact, mobile device that was used in an airborne radar system that was 100 times more powerful than anything that had come before it. The British shared that with the United States. When one of the American military officers saw it for the first time, and fully comprehended what it could be used for, he described the device as having "a giant's strength in a dwarf's arm."

This invention had a huge impact on tactics, strategies, and outcomes. This single piece of powerful, yet mobile technology might even have changed the outcome of the war effort.

Small, powerful, energy efficient processors, such as Intel's Atom and now ARM, are being coupled with versatile software environments, such as Ubuntu and Linux, to create unimagined new futures. These highly mobile systems pack a "giant's strength in a dwarf's arm" — and they are now available to millions of users.

Summary

By reading this book, you're at the front lines of exciting new developments along the Mobile Linux frontier. Your development efforts will be creating the future and a new generation of improvements.

2

The Development Environment

The development process for a mobile device tends to be different from the normal process that you might use when you develop on a workstation or server. Often you'll need to generate a complete OS image from scratch in a working environment, which you'll then install on the target device.

In the case of the Ubuntu Mobile project, a complete set of built packages is available to the developer, so many of the complexities that are typically associated with existing build-your-own-OS-from-scratch projects are not an issue. Various tools bridge the remaining gap between Ubuntu Mobile development and normal Ubuntu development.

This chapter shows different ways you can access the development environment that you'll need in order to begin development. When developing for a mobile device, which perhaps has a different architecture from your workstation — for example, an i386 desktop and a Low Power Intel Architecture (LPIA) device — it becomes necessary to separate the host (sometimes called the source) from the guest (sometimes called the target) environment. The most common way today to do this is to use virtualization.

Virtualization completely separates an operating system from the underlying platform resources — different technologies can be used to make this happen. For the Hardy release, Ubuntu chose to focus virtualization efforts on KVM. Kernel Virtual Machine (KVM) is a patch to the Linux kernel that enables guest operating systems to sit directly on the host hardware. KVM also employs Quick Emulator (QEMU) to turn the Linux Kernel itself into a Hypervisor.

> A Hypervisor is software that allows multiple operating systems to run on a host computer concurrently. QEMU is virtualization software that emulates hardware.

KVM, then, is intended for systems where the processor has hardware support for virtualization and it has the added advantage of falling back to QEMU if this is not available. By sitting on the

Chapter 2: The Development Environment

Linux kernel, guest OSes appear as Linux processes and can be managed just like any other Linux application.

KVM does not, however, currently offer a GUI that could help a developer new to Ubuntu Mobile get quickly up-to-speed with development. VirtualBox from Sun Microsystems, Inc. is a virtualization solution available under the GPL that has a nice front end and is simple to get up-and-running. Some examples are described in this chapter using both VirtualBox and KVM. Finally, for maximum control of the development environment, you learn how to create an image from scratch. This method offers more opportunity for customization, and is the preferred choice for more experienced developers.

Getting Started

Begin by downloading the Jaunty Ubuntu MID release from `http://cdimage.ubuntu.com/releases/jaunty/release/` and save it into a folder on your hard drive.

All of the techniques demonstrated in this chapter (apart from the section on ARM) use the Ubuntu MID release; however, they were also tested and work with the Ubuntu Netbook Remix release. Download this release if you are targeting a netbook.

Getting Familiar with the Ubuntu Mobile Environment

This section covers the Ubuntu Mobile environment.

VirtualBox

Sun xVM VirtualBox is a collection of virtual machine tools that allow a developer to run multiple operating system images at the same time. To see this in action, first install the software.

```
$ sudo apt-get install virtualbox-ose
```

Virtualbox-ose is located in the universe repository. To enable this community-maintained software repository, go to System ⇨ Administration ⇨ Software Sources and select the universe box.

When it is installed, it is possible to convert the image that you downloaded in the "Getting Started" section earlier so it can be used in VirtualBox.

To do this, run the command (substitute the name of the .img file with the one that you downloaded):

```
$ vboxmanage convertdd ubuntu-9.04-mid-lpia.img ubuntu-9.04-beta-mid-lpia.vdi
```

This converts ubuntu-9.04-beta-mid-lpia.img to the .vdi format so it can be used as a "hard disk" in VirtualBox (this is explained in more detail later).

Chapter 2: The Development Environment

Now start VirtualBox by selecting Application ➪ Accessories ➪ VirtualBox OSE. You are presented with the interface that's shown in Figure 2-1.

Figure 2-1

Click New, which starts the setup wizard. Proceed by selecting Next, which will present the VM Name and OS Type dialog. Complete this dialog as shown in Figure 2-2 and then click Next.

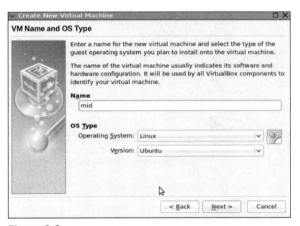

Figure 2-2

On the next screen, choose the maximum amount of memory allowed for your virtual machine (the Next button cannot be clicked when the maximum amount of memory allowed by the hardware has been exceeded). The next screen, shown in Figure 2-3, is for the hard disk.

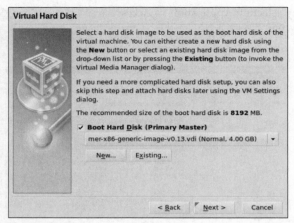

Figure 2-3

Click New and then Next. Select the Dynamically expanding storage option and click Next again. The disk image by default will be saved in ~/.VirtualBox/HardDisks/<name>, so in this example the hard disk shown in Figure 2-4 will be saved as mid.vdi. in the /home/ian/.VirtualBox/HardDisks directory.

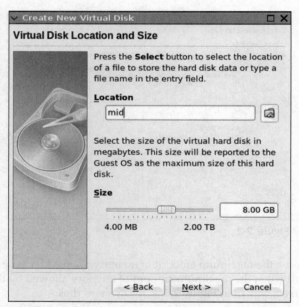

Figure 2-4

Chapter 2: The Development Environment

Choose Next and then Finish and you should see Figure 2-5.

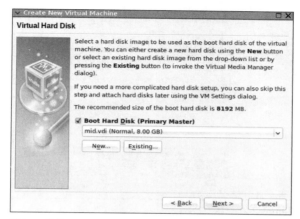

Figure 2-5

This Primary Master hard disk has 8GB of space allowed to it. Select Next and then Finish and this will create the new mid virtual machine shown in Figure 2-6.

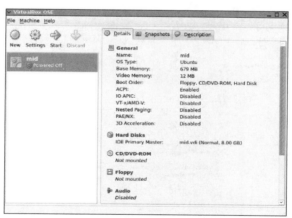

Figure 2-6

Now it is necessary to install Ubuntu MID onto this hard drive so that software can be developed and tested. Click the Settings button in Figure 2-6 and the settings for the virtual machine appear, as shown in Figure 2-7.

15

Chapter 2: The Development Environment

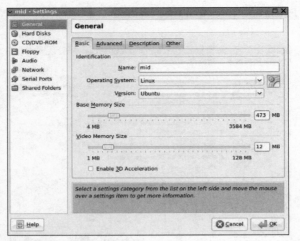

Figure 2-7

Choose Hard Disks, which brings up another screen, as shown in Figure 2-8.

Figure 2-8

It is necessary to add another hard disk into the virtual machine (one that contains the Ubuntu MID operating system). To do this, select the image of the disk with an addition sign on the right hand side and then click the folder image. This brings up the screen shown in Figure 2-9.

Chapter 2: The Development Environment

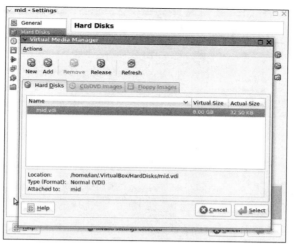

Figure 2-9

Select Add and browse the folder that contains the converted image that was created with the command

```
$ vboxmanage convertdd ubuntu-9.04-mid-lpia.img ubuntu-9.04-beta-mid-lpia.vdi
```

from earlier in this section. Select the converted image and the screen should look like Figure 2-10.

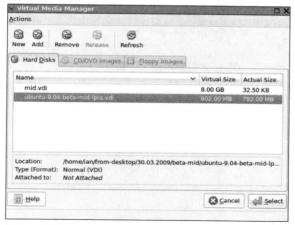

Figure 2-10

Click Select to return to the Hard Disks screen (see Figure 2-11). The ubuntu-9.04-beta-mid-lpia.vdi that was added must be set as the IDE Primary Slave.

Chapter 2: The Development Environment

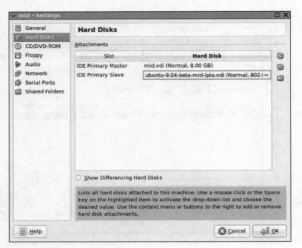

Figure 2-11

Click OK and you will return to the VirtualBox management screen that you saw in Figure 2-6 previously.

It is now possible to start the virtual machine by clicking the Start arrow.

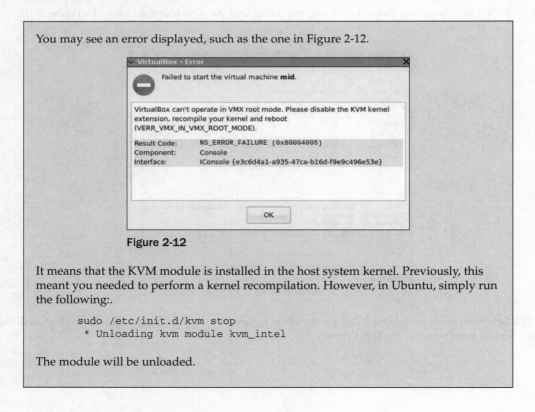

You may see an error displayed, such as the one in Figure 2-12.

Figure 2-12

It means that the KVM module is installed in the host system kernel. Previously, this meant you needed to perform a kernel recompilation. However, in Ubuntu, simply run the following:.

```
sudo /etc/init.d/kvm stop
 * Unloading kvm module kvm_intel
```

The module will be unloaded.

Chapter 2: The Development Environment

As the virtual machine boots, press the F12 key (this will allow selection of the boot device) and you will see the screen shown in Figure 2-13.

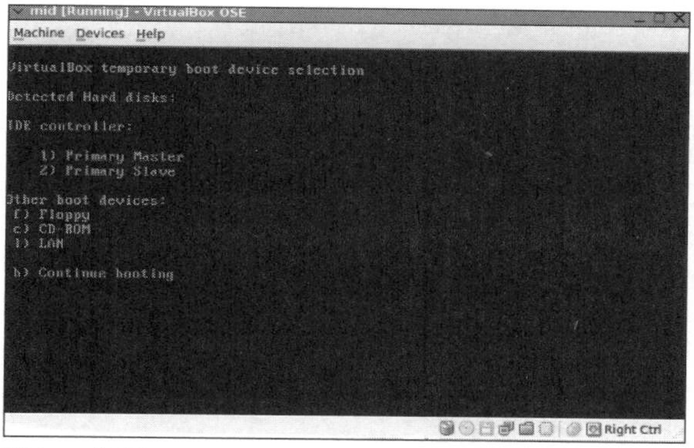

Figure 2-13

Choose option number 2, which is the primary slave (the ubuntu-9.04-mid-lpia.vdi hard disk), to boot. Then choose the option to try out Ubuntu MID and it will boot into the screen shown in Figure 2-14.

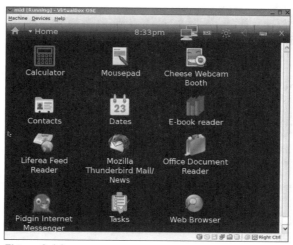

Figure 2-14

In the virtual machine, the top bar of the Ubuntu MID interface contains a menu that allows the user to change the group of programs that are shown. Click the Home drop-down list and choose Preferences. Click the Install icon to install. The system will be installed on mid.vdi, which you created earlier and which is represented in the installer as a hard disk, as shown in Figure 2-15.

Chapter 2: The Development Environment

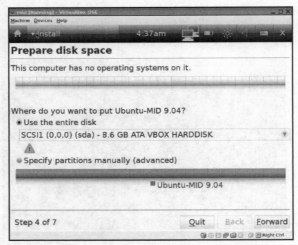

Figure 2-15

When the installation has finished, it is necessary to reboot, as shown in Figure 2-16.

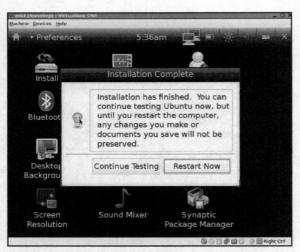

Figure 2-16

It is possible to use hardware virtualization in VirtualBox. This makes VirtualBox similar to KVM in that the hardware is used to speed up the virtualization process. To enable this, first check that your processor has support for this by running:

```
$ egrep --color=auto '(vmx|svm)' /proc/cpuinfo
```

Something should be printed in the terminal if processor support is available. Then go to the General Settings in the VirtualBox Management screen and choose the Advanced tab. Select the Enable VT - x/AMD - V checkbox. Doing this boosts the speed of the virtual machine considerably.

Chapter 2: The Development Environment

KVM/QEMU

KVM (for Kernel Virtual Machine) is a full virtualization solution for Linux. It consists of a loadable kernel module, kvm.ko, that provides the core virtualization infrastructure and a processor-specific module, kvm-intel.ko or kvm-amd.ko. It therefore relies on Intel VT or AMD-V hardware and it is a full virtualization for Linux.

KVM is intended for systems where the processor has hardware support for virtualization. To check this on a system, run the following command:

```
$ egrep --color=auto '(vmx|svm)' /proc/cpuinfo
```

If vmx or svm is displayed in the output, the hardware (CPU) will support KVM. If nothing is printed in the terminal, it means that the CPU doesn't support hardware virtualization.

> *Often hardware support needs to be specifically enabled in the BIOS — as was the case on a Lenovo Thinkpad X60s.*

If nothing is printed, KVM falls back to the slower QEMU-based software virtualization. QEMU is also used for virtualization of other architectures such as ARM. This is explained later in this chapter.

Using QEMU

To use QEMU, install the necessary packages:

```
$ sudo apt-get install qemu kqemu-common kqemu-source
```

Next, add your user name to the kqemu group. Adding your user name to the kqemu group means that you do not need to run QEMU as the root user.

> *kqemu is the name of the module that is used to accelerate QEMU.*

To add your user name, run:

```
$ sudo adduser <username> kqemu
```

To make these changes effective, it is necessary to log out and then log back in again.

To test QEMU, go to the folder that contains the image that you downloaded from http://cdimage.ubuntu.com and run the following:

```
$ qemu -localtime -m 384 -boot c ubuntu-9.04-beta-mid-lpia.img
```

> *Substitute the .img name for the one that you downloaded, if different from the one above.*

To enable the kqemu acceleration module, it is necessary to pass the parameter -kernel-kqemu like this:

```
$ qemu -kernel-kqemu -localtime -m 384 -boot c ubuntu-9.04-beta-mid-lpia.img
```

Chapter 2: The Development Environment

If you get the error message "MP BIOS BUG: MP-BIOS bug: 8254 timer not connected to IO-APIC," it is necessary to boot with the noapic option enabled. To do this, choose F12 when the screen appears and choose 2 to boot from the hard disk. Then choose F6. This will display the menu that's shown in Figure 2-17. Select the noapic option.

Figure 2-17

ARM on QEMU

Before installing Ubuntu ARM on real hardware, it is better to try it with an emulator first. Working in this way also allows applications to be built and packaged using a native compiler rather than cross-compiling.

> *A cross compiler is a compiler that is capable of creating executable code for an architecture other than the one on which the compiler is run. Creating ARM binaries on an i386 environment is one example.*

To do this for Ubuntu Jaunty, download the script from http://people.ubuntu.com/~ogra/arm/build-arm-rootfs. Make it executable:

```
$ chmod +x build-arm-rootfs
```

Note that QEMU and debootstrap need to be installed.

Run the following script (this command takes a long time):

```
sudo ./build-arm-rootfs -f pumd-arm -l <user> -p <password> --notarball
```

> Take a note of this username and password in the command above because they are the username and password used to login into the guest machine.

Chapter 2: The Development Environment

This will create an .img in the folder. Next, download a kernel that works with ARM from `http://people.ubuntu.com/~ogra/arm/qemu/kernel/vmlinuz-2.6.28-versatile` and then run qemu-system-arm, adding both the .img and the kernel as parameters:

```
qemu-system-arm -M versatilepb -kernel ./vmlinuz-2.6.28-versatile
-hda qemu-armel-200904291613.img -m 256 -append "root=/dev/sda mem=256M ro"
```

You need to substitute the .img name for the one the script created; the build-arm-rootfs script is date-based, so running it will result in something like qemu-armel-YYYYMMDDHHMMSS.img.

This will execute QEMU and allow a login with the user and password specified during the .img creation step. The native architecture can be verified, as shown in Figure 2-18.

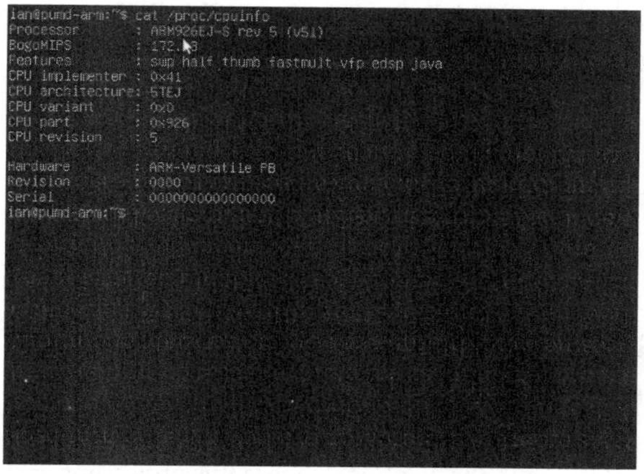

Figure 2-18

Using KVM

If your hardware supports KVM, the next step is to install the necessary package:

```
$ sudo apt-get install kvm
```

Next, add your user to the `kvm` group. Adding your user name means that you do not need to run KVM as the root user. To do this, run the following:

```
$ sudo adduser <username> kvm
```

It is necessary to reboot the virtual machine after adding a new user. Next, run the .img in KVM using the following:

```
$ kvm -localtime -m 512 -boot c ubuntu-9.04-beta-mid-lpia.img
```

Chapter 2: The Development Environment

To install the Ubuntu MID image, a similar process to the one for the VirtualBox installation will be followed. Set up the .img as the second hard disk so you can install on the first disk. You might think of doing something like this:

```
$ kvm -localtime -m 512 -hda installed-mid.img -hdb ubuntu-9.04-beta-mid-lpia.img
```

However, the preceding command will *not* work! Besides the fact that installed-mid.img does not exist, the Bochs BIOS that's used in QEMU and KVM doesn't allow for booting on the second hard disk. As a workaround, the Ubuntu Mobile Team provides a small (less than 400KB) ISO image from which the second hard disk can be chosen to boot. This is available at http://people.ubuntu.com/~lool/isolinux.iso.

The script to generate the preceding ISO is available using bzr (see Appendix A), which you can find at lp:~ubuntu-mobile-dev/ubuntu-mobile/mobile-scripts.

Some of the scripts are discussed in greater detail later in the chapter.

Boot KVM with the following:

```
$ kvm -m 512 -cdrom isolinux.iso -hda installed-mid.img -hdb ubuntu-9.04-mid-lpia.img
```

installed-mid.img in the command above is created using the following:

```
$ qemu-img create installed-mid.img 4G
```

Here 4G stands for 4GB of virtual hard disk space. Because the installed-mid.img is blank, it will boot using the -cdrom option specified in the above command When it does this, isolinux.iso will offer to boot from the second hard disk with the 2 option. Choose this option and install to the first hard disk (where grub will be installed as well).

The install itself is the same as the VirtualBox install; on the virtual machine, the top bar of the Ubuntu MID interface contains a menu that allows you to change the group of programs that is shown. Click the Home drop-down menu and choose Preferences. You'll see an Install icon. Click the icon to install.

Try to use the maximum screen resolution possible on the host because KVM shows the guest OS at the same resolution as the host. At lower resolutions, the buttons on the install will be hidden, and you may need to remove panels on the host to see them! This should be fixed in the final Jaunty MID release.

Reboot the virtual machine when the install process has finished and KVM will start from the now installed hda.img.

It is possible to drop -cdrom and -hdb flags from the KVM command line after the install:

```
$ kvm -localtime -m 512 -hda installed-mid.img
```

Chapter 2: The Development Environment

NETWORKING

The following commands are necessary for networking in both VirtualBox and KVM. First, on the guest computer, log in with the user account created previously, open a terminal and install openssh-server:

```
$ sudo apt-get install openssh-server
```

Then add a user:

```
$ sudo adduser <username>
```

This sets up the basic tools that are required for networking.

If you would like full networking between the host and guest, see the "Advanced Networking on VirtualBox and KVM/QEMU" section.

Networking in VirtualBox

On the host computer, use the VBoxManage command to open up a TCP service port which will be used for SSH. Then, run the following:

```
$ VBoxManage setextradata mid ↵
"VBoxInternal/Devices/pcnet/0/LUN#0/Config/ssh/HostPort" 2222
$ VBoxManage setextradata mid ↵
"VBoxInternal/Devices/pcnet/0/LUN#0/Config/ssh/GuestPort" 22
$ VBoxManage setextradata mid ↵
"VBoxInternal/Devices/pcnet/0/LUN#0/Config/ssh/Protocol" TCP
```

The three commands above should be on single lines. The VBoxManage command allows you to get more information from your virtual machine and it also allows you to perform actions and configuration fine tuning. For example, the name of the virtual machine - mid in the above example - was obtained by running:

```
$ VBoxManage list vms | grep Name
```

Completely shutdown the guest computer using the VirtualBox management console. On the host computer, enter the following:

```
$ ssh -l <username> -p 2222 localhost
```

You now have SSH access to the MID guest:

```
$ ssh -l <username> -p 2222 localhost
```

The HostPort (2222 in this case) must be greater than or equal to 1024 because listening on ports 0–1023 requires root permissions.

25

Chapter 2: The Development Environment

Networking in KVM/QEMU

The default virtual network configuration with KVM/QEMU is usermode networking. Usermode networking makes the virtual machine behave as if it were behind a firewall that blocks all incoming connections (the only "pingable" address is the default DHCP server/firewall on 10.0.2.2). What this means in practice is that, by default, the guest computer has access to the outside network, but you cannot access the host machine.

To override usermode networking in order to enable access to the host, start KVM like this:

```
$ kvm -hda installed-mid.img -m 1024M -redir tcp:2222::22
```

On the host, enter the following:

```
$ ssh -l <username> -p 2222 localhost
```

Advanced Networking on VirtualBox and KVM/QEMU

If it is a requirement to have the virtual machine with full network support or if many virtual machines will be running on the host, it is necessary to set up bridging and a switch to connect the host and the guest.

The following configuration can be used for advanced networking with both VirtualBox and KVM/QEMU.

First, install `vde2` and `dnsmasq` on the host:

```
$ sudo apt-get install vde2 dnsmasq
```

VDE is a virtual switch that can connect multiple virtual machines and dnsmasq is a small caching DNS proxy and DHCP/TFTP server.

Next, add the following lines to /etc/network/interfaces on the host. The name qtap0 in the following code snippet can be called anything you like, as it is a virtual connection and does not really exist on the hardware. Likewise, the 172.12.120.1 address has been chosen because it is uncommon and less likely to conflict with any other network to which your computer is connected. You can choose anything here:

```
auto qtap0
iface qtap0 inet static
    address 172.12.120.1
    netmask 255.255.255.0
    vde2-switch -
    post-up /etc/init.d/dnsmasq start
    # Allowing access to the network
    post-up /sbin/sysctl -w net.ipv4.ip_forward=1
    # assuming that the interface on the host connected to the network is eth0
    post-up /sbin/iptables -t nat -A POSTROUTING -o wlan0 -j MASQUERADE
```

Chapter 2: The Development Environment

```
# Closing access to the network
post-down /etc/init.d/dnsmasq stop
post-down /sbin/sysctl -w net.ipv4.ip_forward=0
post-down /sbin/iptables -t nat -F
```

Following the netmask are actions such as pre-up, which will be undertaken during the network configuration and before the interface is enabled. To see more options that can be added into this file look at the following:

```
$ man interfaces
```

Next, you set up the DNS and caching server. To do this, edit the dnsmasq config file that's found in /etc/dnsmasq.conf and find the interface (line 85):

```
#interface=
```

Change this to

```
interface=qtap0
```

or to whatever name you chose for this interface. Then look for line 147:

```
#dhcp-range=192.168.0.50,192.168.0.150,12h
```

Change it as follows:

```
dhcp-range=172.12.120.100,172.12.120.150,12h
```

Save the file and restart the network on the host machine:

```
sudo /etc/init.d/networking restart

 * Reconfiguring network interfaces...
 * Restarting DNS forwarder and DHCP server dnsmasq

   ...done.


                                                                           [ OK ]
```

Now when you run ifconfig, you can see the new network interface:

```
$ ifconfig
qtap0     Link encap:Ethernet  HWaddr d2:3d:a9:96:88:36
          inet addr:172.12.120.1  Bcast:172.12.120.255  Mask:255.255.255.0
          inet6 addr: fe80::d03d:a9ff:fe96:8836/64 Scope:Link
          UP BROADCAST RUNNING MULTICAST  MTU:1500  Metric:1
          RX packets:0 errors:0 dropped:0 overruns:0 frame:0
          TX packets:29 errors:0 dropped:0 overruns:0 carrier:0
          collisions:0 txqueuelen:500
          RX bytes:0 (0.0 B)  TX bytes:5836 (5.8 KB)
```

Chapter 2: The Development Environment

It is also necessary to add your user name to the host machine and to the vde2-net group, and log out and in again:

```
$ sudo adduser username vde2-net
$ logout
```

Using the Bridge in VirtualBox

In the VirtualBox management console, go to the settings for the virtual machine and choose Network. The adapter needs to be attached to the interface qtap0, which was created previously. In the drop-down menu, choose Host Interface and on the bottom tab select qtap0, as shown in Figure 2-19.

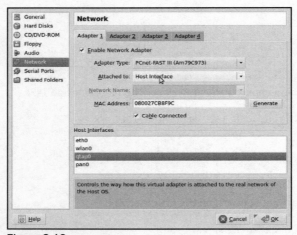

Figure 2-19

Now when the virtual machine starts, it is using the interface qtap0. Find the IP address of the guest machine by looking at the dnsmasq.leases file found at /var/lib/misc/dnsmasq.leases, or by opening a terminal on the guest and running $ifconfig. After this, it is possible to ssh into the guest as seen in the following example:

```
$ ssh 172.12.120.115  -l <username>
```

Files can be copied from the host machine to the guest machine, as follows:

```
$ scp pumd.tar.gz <username>@172.12.120.115:/home/<username>/
```

The guest machine can now ping the host machine.

Chapter 2: The Development Environment

Using the Bridge in KVM

In order to use this new interface, start KVM using the command `vdekvm`:

```
$ vdekvm -localtime -m 512 -net vde,vlan=0
-net nic,model=rtl8139,vlan=0,macaddr=00:16:3e:12:34:56 -hda installed-mid.img
```

The `macaddr` from the preceding command can be any mac address; this is just a made-up one.

When using only one virtual machine, it is not necessary to specify the mac address. However, for multiple virtual machines, it becomes important. Selecting an address from the range 00:16:3e:xx:xx:xx is advisable as this range has been assigned to virtualization.

In order to use the interface with kqemu in QEMU, use the following command:

```
$ vdeqemu -kernel-kqemu -hda installed-mid.img
-net vde,vlan=0,sock=/var/run/vde2/qtap0.ctl -net nic,model=rtl8139,vlan=0,macaddr=
52:54:00:12:34:56 -m 1024 -no-acpi
```

On the host, enter

```
$ ssh 172.12.120.115  -l <username>
```

to ssh on the guest machine. The IP above is just an example.

Files can be copied like this:

```
$ scp pumd.tar.gz <username>@172.12.120.1:/home/<username>/
```

The guest machine can now ping the host machine.

Sharing Files Between Guests and Host

If you followed the simple networking procedure in the "Networking in VirtualBox" or "Networking in KVM/QEMU" sections, it is possible to copy files from the host computer to the guest computer by typing the following:

```
$ scp -P 2222 <file> username@localhost:/home/<username>/
```

However, it is also possible to use sshfs to copy files. This is done by installing sshfs in the host

```
$ sudo apt-get install sshfs
```

and then by running the command on the host:

```
$ sshfs -p 2222 username@localhost:/home/<username>/
```

If you want to automatically mount the user's home directory, add the following line to /etc/fstab

```
sshfs#username@localhost:/home/<username>  /media/guest-home fuse user,noauto 0 0
```

then add the user currently logged in on the host to the fuse group:

```
$ adduser <host username> fuse
```

Logout and login again to apply the changes. Now you can mount the virtual machines home/user directory by typing:

```
$ mount /media/guest-home
```

Now files can be copied from the host to the guest and vice versa through this mount point.

Everything mentioned so far in this section works for simple networking. If you followed the "Advanced Networking on VirtualBox and KVM/QEMU" section's instructions, then a slightly different procedure is required.

Sharing Files Between Guests and Host with Advanced Networking

If you want to copy files using scp from the guest to the host you can use:

```
$ scp <file> username@172.12.120.1:/home/<username>/
```

To copy files from the host to the guest, first get the guest IP address and then run for example on the host:

```
$ scp <file> username@172.12.120.100:/home/<username>/
```

The sshfs solution presented above also works when using advanced networking. First install sshfs in the host

```
$ sudo apt-get install sshfs
```

and then run the following command on the guest:

```
$ sshfs username@172.12.120.100:/home/<username>/ ~/devel
```

Add the currently logged in user on the host to the fuse group:

```
$ adduser <guest username> fuse
```

Logout and login again to apply the changes. Now you can mount the host's home/user directory by adding the following line to /etc/fstab

```
sshfs#username@172.12.120.1:/home/<username> /media/host-home fuse user,noauto 0 0
```

and mounting:

```
$ mount /media/host-home
```

If you want to mount the user's home directory on startup, add the following line to /etc/fstab:

```
sshfs#username@172.12.120.1:/home/username /media/host-home fuse defaults,allow_
other 0 0
```

Chapter 2: The Development Environment

If you do not want a password prompt every time you copy files with scp or with the mount solutions explained above you need to set up passwordless ssh keys. On the host enter:

```
$ ssh-keygen -t rsa
```

Just hit Enter when asked for a password. Then run the following on the host:

```
$ ssh-copy-id -i"/home/<host username>/.ssh/id_rsa.pub -p 2222 <username>@localhost"
```

Now you are able to ssh, copy and mount without being asked for a password.

Building Your Own Virtual Image

Rather than using a pre-rolled image from cdimage.ubuntu.com, it is possible to build a virtual machine image. To do this, install a tool called ubuntu-vm-builder:

```
$ sudo apt-get install ubuntu-vm-builder
```

Then run it like this:

```
$ sudo ubuntu-vm-builder kvm jaunty --mem=512 --user <username> --pass <password> --addpkg openssh-server
```

This creates a minimal KVM virtual machine for Jaunty, adds a user with a password, and adds the openssh-server package to the image. Creating a virtual image this way can be useful in order to provide a clean testing environment. It can also be fast if the preceding command is combined with a local repository mirror such as:

```
$ sudo ubuntu-vm-builder kvm jaunty --mem=512 --user <username> --pass <password> --addpkg openssh-server --mirror http://localhost:9999
```

This command creates a folder with two files in it: disk0.qcow2, which is the QEMU image, and a script called run.sh. Execute the script like this:

```
$ ./run.sh
```

This passes the qcow image as a parameter to QEMU.

Working with Images

The Ubuntu-Mobile team has made some scripts available, which you'll find in the mobile-scripts package at `https://code.edge.launchpad.net/~ubuntu-mobile-dev/ubuntu-mobile/mobile-scripts`. These scripts are used by the Ubuntu Mobile developers when they're working with images. To get the scripts first, install Bazaar, which is a source code management tool that is used by Ubuntu and many other software projects:

```
$ sudo apt-get install bzr
```

Chapter 2: The Development Environment

Next, use Bazaar to check out the scripts:

```
$ bzr branch lp:~ubuntu-mobile-dev/ubuntu-mobile/mobile-scripts
```

This will create a folder called mobile-scripts with a subfolder called image-tools. The scripts that will be used are:

- create-image.sh
- edit_squashfs.sh
- grow_image.sh

Building Your Own Image

Actual images like the ones downloaded from cdimage.ubuntu.com are made up of a filesystem, which in embedded development is usually squashfs. squashfs is a compressed read-only filesystem that is useful when working with constrained hard drives and memory systems commonly found in embedded environments.

For Ubuntu Mobile this squashfs filesystem will contain an install (for example, an install of Ubuntu MID) as well as a kernel, an initramfs (a special instance of a temporary filesystem), and a syslinux install (syslinux is a boot loader).

To create an image then, it is necessary to first create a squashfs filesystem. To do this install livecd-rootfs:

```
$ sudo apt-get install livecd-rootfs
```

and then run as root one of the livecd-rootfs scripts called livecd.sh:

```
$ sudo livecd.sh -d karmic -a lpia ubuntu-mid
```

The script livecd.sh is installed into the /usr/sbin/ directory, which is in the $PATH. This means the script can be called from any directory without the user supplying the full path to it. The command above was run on Ubuntu karmic.

This will download and then build the squashfs filesystem for LPIA. It can take some time to download, however. Using a local mirror can help here:

```
$ sudo livecd.sh -d karmic -a lpia -m http://localhost:9999 ubuntu-mid
```

Creating a local mirror is explained in Chapter 11.

When the download is complete, the following files and folders will be created:

- chroot-livecd
- livecd.ubuntu-mid.manifest
- livecd.ubuntu-mid.manifest-desktop

Chapter 2: The Development Environment

- livecd.ubuntu-mid.sort
- livecd.ubuntu-mid.squashfs
- livecd.ubuntu-mid.initrd
- livecd.ubuntu-mid.initrd-lpia
- livecd.ubuntu-mid.kernel
- livecd.ubuntu-mid.kernel-lpia

Now it is possible to use one of the scripts (create-image.sh) from mobile-scripts to actually create the image.

Make sure the script is in the same folder as the files and folders mentioned and that it has the executable permission bit set:

```
$ chmod +x create-image.sh
$ ./create-image.sh ubuntu-mid 40
```

where ubuntu-mid (LPIA) is the distribution to be built and 40 is the extra space in MB to be allocated in the image for installing extra packages and other customizations that may be required.

The edit_squashfs.sh script (shown next) can also be used to alter the size of an image.

The built image can be tested now in KVM/QEMU:

```
$ kvm -localtime -m 512 -hda ubuntu-mid.img
```

Installing Applications inside the Image

The script edit_squash.fs from mobile-scripts spawns a root shell inside the image that was just built in the previous section. It is then possible to run `apt-get update` and install packages. On exiting the shell with Ctrl+D or the `exit` command, the script will offer to re-roll the squashfs for you.

The script is used like this:

```
$ sudo ./edit_squashfs.sh ubuntu-mid.img
```

It is always a good idea to run `apt-get clean` when inside the chroot and before exiting to clean up space in the /var/cache/apt/archives directory. Indeed the preceding script does this. This is because space is at a premium. Another idea is also to install deborphan and/or gtkorphan in order to remove unused programs and libraries that are taking up space (`apt-get install deborphan`).

The space available to install new applications/packages is what was specified in the `create-image.sh` command (40MB). If this needs to be increased, use the grow_image.sh script discussed next.

Increasing a Downloaded Image Size

To increase the size of the downloaded image (perhaps to install more applications) use the grow_image.sh script.

This script was created to satisfy a potential use case whereby an OEM wants to modify the applications on a device and maybe the theme but does not want to modify the kernel or the seeds.

Seeds are discussed in more detail in Chapter 11.

You use the script like this:

```
$ ./grow_image.sh ubuntu-mid.img  50
```

This creates a new image with the size specified (50MB) and then it is possible to run the edit_squashfs.sh script on this to make the changes necessary and then re-roll the image.

Summary

Mobile development can be challenging. You need to use your workstation to develop for a device with perhaps a different processor, as is the case with Ubuntu ARM or with a different operating system.

Creating an image (or downloading one from the archives) and then running it in VirtualBox or KVM/QEMU will provide an isolated environment. This environment can then be used for software development, software testing, or application packaging.

Power Management

Mobile devices are now equipped with cameras, various radio transmitters, advanced multimedia, and other features — all of which consume a lot of power. At the same time, more power-hungry applications that utilize these features are being provided. Furthermore, user demands for battery life seem exponential. Because a battery life of six hours is now seen as the absolute minimum necessary when away from an energy source, it is important for you as a developer to try and work out how to conserve energy as much as possible.

It is also becoming increasingly important for users to control their mobile devices' power usage. Not only should users be able to specify power usage, but they should also be able to override the system defaults if they choose.

This chapter explains the various tools and packages that are available for power management on Ubuntu Mobile. It includes the existing ones that are based on HAL and the newer ones that are based on DeviceKit-power. The chapter also shows the results of several power tests, and suggests techniques for increasing battery life on mobile devices.

Introduction

PC-based chips already have a well-established code base in Linux, with existing hardware interfaces such as ACPI and HAL; and as they have evolved from desktops into smaller devices many software packages have been created to optimize power. Each of these packages can be thought of as a power micro-policy controller for a specific subsystem on a Linux-based platform. For instance, `cpufreq` manages power for the CPU subsystem and the `iwconfig` interface manages device-specific power for the Network (WLAN) subsystem.

Chapter 3: Power Management

It would be nice if there were a central way to manage power information. From Ubuntu Jaunty there is a new daemon, which can be installed from the universe repositories, that provides such a service. It moves power management and profiling to an interface that is more focused on the entire power management system than anything that existed before. It is called DeviceKit-power and it is accessible through the D-Bus interface org.freedesktop.DeviceKit.Power.

For more information about D-Bus, see Chapter 4.

Power Saving States

The common power (saving) states are awake, standby, suspend, and hibernate:

- ❑ During the *awake* state all components of the device are running.
- ❑ During *standby*, the CPU keeps running your programs, but some components such as the hard disk, may be turned off. When the mouse or keyboard is touched, the hard disk is accessed by software and the device quickly wakes up.

In both the awake and the standby states, the speed of the CPU may be throttled down if it's not in use.

- ❑ During *suspend*, however, the CPU is always stopped. In modern devices all other components (except the RAM memory) can be turned off. The RAM will hold the "state" of the device.
- ❑ During *hibernation* the state is written to hard disk and the whole device is turned off.

Power Management Packages

If the kernel is compiled with power management packages (the default in Ubuntu), both `apm` and `acpi` will be available.

To see the default kernel configuration, run

```
$ vi /boot/config-`uname -r`
```

For more information about kernel compilation, see Chapter 8.

`apm` provides access to battery status information and may help conserve battery power, depending on your laptop and the implementation. It has mainly been superseded by newer power management schemes, such as ACPI.

For example, to use `apm` to put a device to sleep, run the following:

```
$ apm -s
```

`acpi` is an "interface" specification and is provided on Ubuntu by the packages `acpi` and `acpi-support`. `acpi` attempts to replace the old `apm` functionality, and `acpi-support` contains scripts for events such as the lid closure and loss and gain of AC power.

Both `acpi` and `apm` enable hardware events such as the end of battery power or the pressing of a button to be controlled by software, and both provide the control daemons `apmd` and `acpid`, respectively.

The control daemons run the scripts they find in their configuration directory tree under `/etc/acpi`, or `/etc/apm`, respectively.

pm-utils

Another package that comes as default with Ubuntu and which can be used to control power states is `pm-utils`. This provides the `pm-action`, `pm-hibernate`, `pm-suspend`, and `pm-suspend-hybrid` commands. They allow the triggering of hard power management events by software.

These commands will usually be called by the hardware abstraction daemon (`hald`) when triggered to do so by a program on the desktop. Calling them from the command line is also possible, but it is not guaranteed that all programs will keep working as expected.

pm-suspend

Suspend is a state in which most of the device is shut down, except for RAM. In this state, the device still draws power.

It is possible to use `pm-suspend` to enter sleep mode on a device with a command something like the following:

```
$ pm-suspend --quirk-vbestate-restore
```

Testing on an actual device has revealed that this can be quite slow. The actual scripts that `pm-utils` runs for sleep can be found in /usr/lib/pm-utils/sleep.d/. Many of the scripts there are solutions to problems that are currently being fixed in the kernel.

Always make sure the kernel is updated or alternatively roll a custom kernel. See Chapter 8 for more details.

pm-hibernate

During hibernate mode, the state of the system is saved to disk, and the system is fully powered off, except perhaps for a very low power state on — for example — an Ethernet card to enable wake-on-lan.

pm-suspend-hybrid

Hybrid-suspend is the process where first the state of the system is saved to disk — as with hibernate — but instead of power off, the device goes into a suspended state, which means it can wake up quicker than from normal hibernation. Its advantage over suspend is that it can resume even if the device runs out of power.

You can also set the device to go into high-power and low-power mode; the command `pm-powersave` is used with an additional parameter of `true` or `false`. It basically works the same as the suspend framework.

Chapter 3: Power Management

To do this, use the following:

```
$ sudo pm-powersave true
```

An alternate method to suspend or hibernate an ubuntu-mobile device *without requiring root access* is explained in the "pmi action" section that follows.

pmi action

Available in the universe repositories is a package called powermanagement-interface. This provides an abstracted layer above `acpi`. The main advantage to using this is that it is possible to suspend or hibernate a mobile device without requiring root access.

```
$ sudo apt-get install powermanagement-interface
```

Call the power management interface directly via the following commands:

hibernate:

```
$ /usr/sbin/pmi action hibernate
```

suspend:

```
$ /usr/sbin/pmi action sleep
```

To lock the Gnome session first (i.e., require a password on resumption), issue the following command before issuing the `pmi` command:

```
$ gnome-screensaver-command -lock
```

For example, it is possible to use these commands to create an application entry on Ubuntu Netbook Remix that locks the screen and puts the device into sleep mode.

First make a file containing the following:

```
gnome-screensaver-command -lock; pmi action suspend
```

Then make the file executable with this command:

```
$ chmod +x <file>
```

Go to Preferences ⇨ Main Menu ⇨ Preferences ⇨ New Item. Type **Application in Terminal** in the form and then name the file **Lock and Suspend**. Then choose the file that's been created. Clicking the icon locks and suspends the device.

How pm-utils Works

The main script (pm-action, called via symbolic links as either pm-suspend, pm-hibernate, or pm-suspend-hybrid) executes hook (executable) scripts in an alphabetical order, passing the parameter `suspend`(suspend to RAM) or `hibernate` (suspend to disk) as instructed. Once all the scripts are finished, the device goes to sleep.

Chapter 3: Power Management

After the machine has woken up, all the hook scripts are executed in reverse order with the parameter resume (resume from RAM) or thaw (resume from disk) as instructed.

Note that suspend-hybrid is currently a placeholder and is not completely implemented.

The hooks do various things such as preparing the bootloader or stopping the Bluetooth subsystem.

The hooks for suspend are placed in the following:

```
/usr/lib/pm-utils/sleep.d
/etc/pm/sleep.d
```

The hooks for the power state are placed in the following:

```
/usr/lib/pm-utils/power.d
/etc/pm/power.d
```

Hooks in /etc/pm/ take precedence over those in /usr/lib/pm-utils/, so it is possible for a developer to override the defaults.

An example suspend/resume script for Bluetooth, called 49bluetooth, which is found in /usr/lib/pm-utils/sleep.d/, looks like this:

```
. "${PM_FUNCTIONS}"

[-f /proc/acpi/ibm/bluetooth ] || exit $NA

suspend_bluetooth()
{
        if grep -q enabled /proc/acpi/ibm/bluetooth; then
                savestate ibm_bluetooth enable
                echo disable > /proc/acpi/ibm/bluetooth
        else
                savestate ibm_bluetooth disable
        if
}

resume_bluetooth()
{
        state_exists ibm_bluetooth || return
        restorestate ibm_bluetooth > /proc/acpi/ibm/bluetooth
```

Chapter 3: Power Management

```
        }


    case "$1" in

        hibernate|suspend)

            suspend_bluetooth

            ;;

        thaw|resume)

            resume_bluetooth

            ;;

        *) exit $NA

            ;;

esac
```

Gnome-Power-Manager

Gnome-Power-Manager is a program with a graphical user interface that subscribes itself to power events and acts on them. It shows you the battery status on laptops and dims the screen if running on battery power, for example. It will also shut down or hibernate the computer after some idle time or before the battery runs out if a user is logged in. The software suspend policy of Gnome-Power-Manager is actually handled by the hardware abstraction layer.

Ubuntu-MID does not show the Gnome-Power-Manager applet on the desktop (although it is running in the background) and consequently it has no provision for a graphical shutdown. The power applet code in Appendix D provides such functionality and can be used if this situation arises.

Gnome-Power-Statistics

This application is a graphical viewer which interacts with the Gnome-Power-Manager over D-Bus using the `org.gnome.PowerManager.Statistics` interface.

Figure 3-1 shows the effects of resuming from suspend mode on a netbook and the associated spike in power usage.

Chapter 3: Power Management

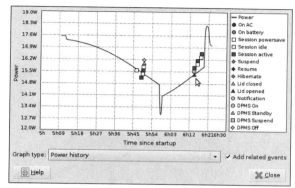

Figure 3-1

Device Kit Power

The DeviceKit set of daemons replaces the core functionality of HAL — the Hardware Abstraction layer. The daemons were written because applications currently need a lot of complicated code to do simple things such as work out the remaining battery life. All of the DeviceKit daemons are system activated and have low memory requirements.

DeviceKit-power also implements a QoS (Quality of Service) interface for latency control (discussed shortly). This is needed to produce a device that uses little power when idle, but at the same time doesn't feel sluggish.

DeviceKit-power provides a D-Bus interface (for more information on D-Bus see Chapter 4) for power sources on the device and to control device-wide power management. Any application can access the org.freedesktop.DeviceKit.Power service on the system message bus.

DeviceKit.Power can be installed using the following:

```
$ sudo apt-get install devicekit-power
```

`Devkit-power-daemon` provides the org.freedesktop.DeviceKit.Power service on the system message bus and `devkit-power` provides a handy command line tool to access the daemon. The tool can be used like this (the following output has been edited for clarity):

```
$ devkit-power --monitor-detail
Monitoring activity from the power daemon. Press Ctrl+C to cancel.
device changed:      /org/freedesktop/DeviceKit/Power/devices/battery_BAT0
    native-path:
/sys/devices/LNXSYSTM:00/device:00/PNP0A08:00/device:01/PNP0C09:00/PNP0C0A:00
/power_supply/BAT0
    vendor:              SONY
    model:               93P5030
    serial:              1656
    power supply:        yes
    updated:             Tue May  5 21:16:40 2009 (0 seconds ago)
```

Chapter 3: Power Management

```
      has history:          yes
      has statistics:       yes
      battery
        present:            yes

        rechargeable:       yes
        state:              discharging
        energy:             74.75 Wh
        energy-empty:       0 Wh
        energy-full:        74.88 Wh
        energy-full-design: 74.88 Wh
        energy-rate:        19.708 W
        voltage:            16.248 V
        time to full:       0 seconds
        time to empty:      3.3 hours
        percentage:         96.1672%
        capacity:           100%
        technology:         lithium-ion
      History (charge):
        1241554960   97.035   unknown
        1241554970   96.955   discharging
        1241554980   96.888   discharging
        1241554990   96.795   discharging
```

The tool shows the output while it's running on battery power. Running the following

```
$ devkit-power --wakeups

Total wakeups per minute: 216

Wakeup sources:

userspace:1 id:16881, interrupts:95.3, cmdline:/usr/lib/firefox-3.0.10/firefox,
details:schedule_hrtimeout_range (hrtimer_wakeup)

userspace:1 id:16901, interrupts:93.2, cmdline:/usr/lib/firefox-3.0.10/firefox,
details:futex_wait (hrtimer_wakeup)

userspace:1 id:6079, interrupts:3.6, cmdline:gnome-terminal, details:schedule_
hrtimeout_range (hrtimer_wakeup)

userspace:0 id:1056, interrupts:3.6, cmdline:modprobe,
details:usb_hcd_poll_rh_status (rh_timer_func)

userspace:1 id:21543, interrupts:3.1, cmdline:xchat,
details:schedule_hrtimeout_range (hrtimer_wakeup)
```

```
userspace:1 id:7617, interrupts:3.1,
cmdline:/usr/lib/openoffice/program/soffice.bin,
 details:schedule_hrtimeout_range (hrtimer_wakeup)

userspace:1 id:17320, interrupts:3.1,
cmdline:/usr/lib/firefox-3.0.10/firefox,
details:schedule_hrtimeout_range (hrtimer_wakeup)

userspace:1 id:5174, interrupts:1.6,
cmdline:avahi-daemon: running [lawrence.local],
details:schedule_hrtimeout_range (hrtimer_wakeup)

userspace:1 id:5397, interrupts:1.0, cmdline:/usr/sbin/apache2,
details:schedule_hrtimeout_range (hrtimer_wakeup)

userspace:1 id:3348, interrupts:1.0, cmdline:/usr/bin/vde_switch,
details:hrtimer_start (it_real_fn)

userspace:1 id:7651, interrupts:1.0,
cmdline:/usr/lib/openoffice/program/soffice.bin,
details:schedule_hrtimeout_range (hrtimer_wakeup)
```

produces Userspace and kernel wakeups (again the output has been edited for clarity). Constant wakeups will adversely affect power consumption as they force the processor to stay "awake" and drain the battery very quickly.

DeviceKit.Power is accessible through the system message D-Bus. To suspend a device, for example, run the following:

```
$ dbus-send --print-reply --system --dest=org.freedesktop.DeviceKit.Power /org/freedesktop/DeviceKit/Power org.freedesktop.DeviceKit.Power.Suspend
```

It returns the following

```
method return sender=:1.161 -> dest=:1.168 reply_serial=2
```

The device enters suspend mode.

The Quality of Service: QoS Interface

To illustrate how you use the QoS interface, an Instant Messenger application needs to request a latency of 200 microseconds.

It was possible to control power from an application before the QoS interface (using the old HAL interface); however, this meant running the application as root.

First, to see the daemon output, run the daemon in verbose mode:

```
$ sudo su
# /usr/lib/devicekit-power/devkit-power-daemon --verbose
```

Chapter 3: Power Management

Pass (in another shell window) the latency request of 200 microseconds (this seems reasonable for an Instant Messenger application, for example):

```
$ dbus-send --print-reply --system --dest=org.freedesktop.DeviceKit.Power
/org/freedesktop/DeviceKit/Power/Policy
org.freedesktop.DeviceKit.Power.QoS.RequestLatency  string:"network" int32:200
boolean:false
```

The output is

```
TI:15:03:48     TH:0x8d4adf8     FI:dkp-qos.c     FN:dkp_qos_request_latency,326


 - Received Qos from ':1.140' (200:1)' saving as #1524452718


TI:15:03:48     TH:0x8d4adf8     FI:dkp-qos.c     FN:dkp_qos_latency_write,180
```

When the IM application disconnects from the system bus, the latency request is automatically cleaned up (the persistent=False setting). This can be set to True and the request is not cleaned up when the application disconnects (the cookie value is preserved across device reboots).

Controlling Radio Transmitters

Constant polling for radio connections of all kinds can use a lot of power.

To try to limit this, device manufacturers have implemented various techniques. On Nokia devices, such as the N810, all network connections are routed through a connection tracker. This makes it possible to disable the network interface until an application tries to access the network, when a new connection can be established quickly. The Power Save Poll protocol (PS-Poll) was developed and helps reduce the amount of time a radio needs to be powered.

Rather than having the radio on all the time, PS-Poll allows the WiFi adapter to notify the access point when it will be powered down. While the radio is powered down, the access point will hold any network packets that would need to be sent to it.

> *Network latencies increase with PS-Poll. This is generally not noticeable when browsing the Web, but in situations where lower network latency is important (for example, online gaming, voice and media streaming), PS-Poll may not be the best solution.*

PS-Poll functions principally with wireless cards that use the ipw2100/ipw2200 driver or the newer iwl3945/ iwl4965 drivers. To set PS-Poll on an ipw2100/ipw2200 card do the following:

```
iwpriv wlan0 set_power 5
```

where wlan0 is the interface name for the network adapter. The number 5 in the iwpriv command is the degree to which the power saving should be enabled:

Chapter 3: Power Management

- ❑ 1 is the lowest number, with the least savings (but also with the lowest added latency).
- ❑ 5 is the highest number.
- ❑ 6 disables the power savings feature again, as in the following:

```
iwpriv eth1 set_power 6
```

The iwl3945/iwl4965 drivers rely on the kernel for configuring the wireless subsystem. A mechanism exists, however, for manually configuring the PS-Poll via a sysfs file attribute `power_level` such as

```
echo 5 > /sys/bus/pci/drivers/iwl4965/*/power_level
```

Substitute iwl3945 for iwl4965, depending on the adapter.

RFKILL

Because there are situations, such as during a flight, when the WiFi radio will not be used at all, it is sensible to turn WiFi off altogether.

To do this, run the following:

```
for i in `find /sys -name "rf_kill" ; do echo 1 > $i ; done
```

To turn the radio back on, run the following in a shell:

```
for i in `find /sys -name "rf_kill" ; do echo 0 > $i ; done
```

It would be a nice touch for an OEM to add an applet for this.

See Chapter 4 for an example of how to write an applet.

Bluetooth

Like WiFi, Bluetooth has a radio transmitter, so it can take quite a bit of power. Unlike WiFi, Bluetooth is rarely used, so it's very possible that the Bluetooth device isn't actually used for anything, and is just consuming battery life.

This is a script to turn Bluetooth on/off as required:

```
#!/bin/bash
if ps -A | grep -c bluetoothd
then
gksudo /etc/init.d/bluetooth stop
sudo hciconfig hci0 down
else
gksudo /etc/init.d/bluetooth start
fi
```

Chapter 3: Power Management

Bluetooth devices that are active on a system can be seen with the `hcitool` *command:*

```
# hciconfig
hci0:   Type: USB
        BD Address: 00:00:00:00:00:00 ACL MTU: 0:0 SCO MTU: 0:0
        DOWN
        RX bytes:0 acl:0 sco:0 events:0 errors:0
        TX bytes:0 acl:0 sco:0 commands:0 errors:0
```

Investigating Power Usage

Finding out about power consumption on Linux often involves the use of `powertop` (`sudo apt-get install powertop`).

Run it inside a target:

```
$ sudo powertop
```

It will show something like this:

```
Wakeups-from-idle per second : 48,0
Power Usage (ACPI Estimate): 7,1 W (2,5 hours)

    Top causes for wakeups:

 21.3% (165.7)       <kernel IPI> : Rescheduling interrupts

 20.9% (162.5)       <interrupt> : uhci_hcd:usb1, yenta, i915@pci:0000:00:02.0

 18.1% (140.6)       <interrupt> : libata

 11.4% ( 88.5)       hildon-desktop : schedule_timeout (process_timeout)

  7.1% ( 55.1)       <interrupt> : extra timer interrupt

  5.4% ( 42.3)       midbrowser : futex_wait (hrtimer_wakeup)
```

The `Wakeups-from-idle per second"` *line is an indicator for how well your device is doing in terms of getting power savings: The lower the number the better.*

`Powertop` also shows some handy tips, like this:

```
Suggestion: increase the VM dirty writeback time from 5.00 to 15 seconds with:

  echo 1500 > /proc/sys/vm/dirty_writeback_centisecs

This wakes the disk up less frequently for background VM activity
```

Chapter 3: Power Management

With `powertop`, it is easy to see that some applications have fairly heavy energy requirements. For example, if your battery life is 5 hours, then running Gaim will consume 1 hour of that! Other notable applications include the midbrowser (Firefox), xorg, and Skype. It should be noted that, especially with Firefox, much work has been done recently to reduce wakeups.

See Chapter 10 for additional power-saving advice.

Battery Testing

This test uses the Phoronix Test Suite, which is discussed in greater detail in Chapter 9.

Preparing to Run the Tests

Before you run any test, it is necessary to disable the screensaver (on Ubuntu Netbook Remix):

```
$ sudo killall gnome-screensaver
$ gconftool-2 --type boolean -s
/apps/gnome_settings_daemon/screensaver/start_screensaver false
```

Also unplug external devices (USB-mouse, keyboard, Ethernet, and so on) and then reboot the device.

Phoronix Test Suite

The Phoronix Test Suite supports monitoring system sensors. The Phoronix Test Suite will automatically monitor selected sensors while each test in the chosen suite is running and at the end will provide the low and high thresholds for each sensor as well as the average. In addition to this, the sensor results are then plotted on line graphs.

The sensor detection and monitoring itself is done through LM-Sensors and the ACPI interface. LM-Sensors is a hardware health monitoring package for Linux, which allows access to information from temperature, voltage, and fan speed sensors. It is used as the default but if LM-Sensors isn't installed, the suite will fall back to using the Advanced Configuration and Power Interface.

To install Phoronix Test Suite and the sensor interfaces, run the following:

```
$ sudo apt-get install lm-sensors libsensors-dev phoronix-test-suite
```

The full list of sensor options can be found by running the following:

```
$ phoronix-test-suite module-info system_monitor
```

The sensors that are actually detected on the current system and their values can be read by running the following:

```
$ MONITOR=all phoronix-test-suite test-module system_monitor
```

Chapter 3: Power Management

The results on an Acer Aspire netbook appear like this:

```
======================================

Starting Module Test Process

======================================


==========================================

Current Sensor Readings:
Battery Power Monitor 10603 Milliwatts
CPU Frequency Monitor: 800.00 Megahertz
CPU Usage Monitor: 14.81 Percent
System Memory Usage Monitor: 454 Megabytes
Swap Memory Usage Monitor: 189 Megabytes
Total Memory Usage Monitor: 644 Megabytes
Elapsed Time: 13 Seconds


==========================================
```

Selecting sensors to monitor is performed through the MONITOR environmental variable. For example, to monitor the battery while the netbook suite runs, the following command would be used:

```
$ MONITOR=all.battery phoronix-test-suite benchmark netbook
```

Battery Comparisons

The device that was used in the previous example was an Acer Aspire netbook with an Intel Atom N270 1.6GHz CPU. It is expected that significant battery gains will be achieved using the Low Power Intel Architecture (LPIA) that is based on the Ubuntu MID release. In order to confirm this, the battery was monitored while the idle test ran for one hour:

```
$ MONITOR=all phoronix-test-suite benchmark idle
```

On both operating systems, the battery was fully charged before beginning the test and the power supply was unplugged before running the test.

After the idle test was completed, the sensor statistics were printed in the standard output:

```
Power Battery Statistics:
Low: 10871.00 Milliwatts
High: 14233.00 Milliwatts

Average: 11779.76 Milliwatts
```

The results are shown using a web browser using various images like Figure 3-2.

Chapter 3: Power Management

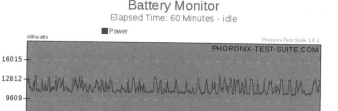

Figure 3-2

On the same Acer netbook, Ubuntu MID was installed and the same test run for one hour. This is the output for Ubuntu MID:

```
Power Battery Statistics:
Low: 9919.00 Milliwatts
High: 13162.00 Milliwatts
Average: 10401.40 Milliwatts
```

The results are shown in Figure 3-3.

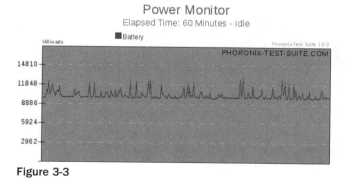

Figure 3-3

The sensor graphs for the preceding test are saved to the following default location: ~/.phoronix-test-suite/module-files/system-monitor/.

The results show that the LPIA-based Ubuntu MID is 11.7 percent more power efficient than the i386 Ubuntu Netbook Remix. The problem with relying on this test result is that it is not comparing similar things. It would be nice to have Ubuntu Netbook Remix running on LPIA and compare that with the i386 version.

Chapter 3: Power Management

Comparing Like-to-Like

First, Ubuntu MID was installed on the Acer and then in a terminal:

```
$ sudo apt-get install ubuntu-netbook-remix-default-settings ubuntu-netbook-remix
```

Next, the MID settings were removed so that the Netbook Remix interface could be used:

```
$ sudo apt-get remove -purge ubuntu-mid-default-settings
```

Finally, the test was run again. This resulted in the following:

```
Power Battery Statistics:

Low: 10250.00 Milliwatts

High: 13111.00 Milliwatts


Average: 10772.64 Milliwatts
```

The results are shown in Figure 3-4.

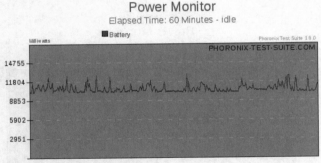

Figure 3-4

These results show that the LPIA version of Ubuntu Netbook Remix is 8.55 percent more power-efficient than the i386 version.

Another interesting point to note is that the LPIA Ubuntu MID release is 3.44 percent more power-efficient than the LPIA version of Ubuntu Netbook Remix. This can be attributed perhaps to the fact that standard Firefox is used on Ubuntu Netbook Remix, whereas midbrowser (which is Firefox-compiled using the LPIA flags) is used on the LPIA version of Ubuntu MID.

Chapter 3: Power Management

Summary

Power management is an essential job on portable computers. At the core of power management is an understanding of how to effectively optimize energy consumption of each system component. This entails studying the different tasks that your system performs, and configuring each component to ensure that its performance is just right for the job.

This is an area of considerable research and one whereby good techniques and innovative solutions can have a direct impact on profit for an OEM. Consumers want more battery life and they are willing to pay for it.

Application Development

The intention of this chapter is not to show how to program in any specific language or how to use any particular graphical toolkit. Rather, the aim is to show, through the use of examples, how to develop small, useful applications that demonstrate key concepts and techniques. Such an approach is, we believe, useful for a developer when more complex applications need to be developed.

Like the author of the Debian New Maintainers Guide, we believe in the Latin saying *Longum iter est per praecepta, breve et efficax per exempla!* (It's a long way by the rules, but short and efficient with examples!) This chapter and the book in general follow this advice.

Mobile Internet Devices, tablets, and netbooks need a desktop framework that can provide a way to create applications with a very consistent look and feel, and run and interface nicely using the restricted resources found on mobile devices. In addition, applications should be designed for touchscreen use with finger-friendly navigation. All of these factors provide challenges to a developer.

This chapter begins by discussing issues related to the design and layout of applications on small form factor devices. It then looks at GTK, which is a toolkit for building graphical applications, and briefly discusses other toolkits that can be used in the mobile and embedded space such as EFL. In the context of GNOME Mobile, two application frameworks are prominent; Hildon and Clutter and example applications are provided for each of these. Next it shows the use of D-Bus, a technology which is important for inter-process communication and which is crucial to the new notifications system on Ubuntu.

Finally, all the concepts presented so far in the chapter are brought together in an example application which demonstrates the use of many of the key technologies a developer should know in order to become productive with Ubuntu Mobile.

Chapter 4: Application Development

Ubuntu Mobile Releases

The major Jaunty releases from the Ubuntu Mobile team are Ubuntu MID and Ubuntu Netbook Remix. For karmic, the MID release will be community maintained. There will be a remix of Moblin called Ubuntu Moblin Remix, and there will also be an ARM release.

> *There is also the port to the armel architecture* http://cdimage.ubuntu.com/ports/releases/jaunty/release/ubuntu-9.04-alternate-armel.img. *This first release is based on the Freescale I.MX515 processor, which is a chip that apparently draws very little power and generates very little heat. This allows developers to squeeze longer battery life (up to 8 hours) out of a very thin form factor.*
>
> *Freescale, along with Pegatron, an Original Device Manufacturer originally part of ASUS, are working with Canonical to customize Ubuntu for this architecture. The Ubuntu Port to arm is discussed further in Appendix A.*

The Ubuntu Netbook Remix is a standard Ubuntu desktop along with some specific customizations that enable it to work better on devices with small screens, such as netbooks.

These are the Ubuntu Netbook Remix Launcher, which provides an alternate menu along with a Go Home Applet, which works in conjunction with the launcher to provide access to the desktop. There is also Maximus, which is a daemon to automatically maximize a window along with a Window Picker Applet, which displays open windows as icons on the panel. These tools are pulled together into a new theme (for jaunty) called the Human Netbook Theme.

It is easy to switch between the "classic" and the netbook desktops by installing the desktop switcher package:

```
$ sudo apt-get install desktop-switcher
```

Run the application and it pops up an interface to choose the desktop that's shown in Figure 4-1.

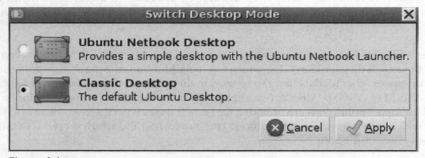

Figure 4-1

This command changes some GConf settings for each desktop — for example:

```
gconftool-2 — set /desktop/gnome/session/required_components_list — list-type=string ["filemanager","panel","windowmanager"]
```

Chapter 4: Application Development

This shows the power of GConf to radically change both the user interface and configuration settings. GConf is discussed later in this chapter.

Developing an application for Ubuntu Netbook Remix is then the same as for a regular desktop. It is with Ubuntu MID that a developer needs to be concerned about some fundamental differences — mainly related to the smaller screen size.

Creating a New Application

OEMs will want to significantly customize the interface of a device as well as the range of default applications presented to their end users "out-of-the-box." This may require applications to be developed if the required functionality does not currently exist in the Ubuntu repositories. In addition to following a user-driven development process and choosing the development methodology (Scrum is highly recommended) there are some other key considerations when developing for mobile devices.

Application Design

The most important thing to consider when developing for mobile devices is the reduction in screen area when compared with a regular desktop. With this in mind during the Ubuntu Developer Sprint in Boston, the mobile team brainstormed some key points to consider when working in this environment:

- Screens and dialogs must fit onscreen. This is normally 800 × 480.
- Applications with multiple screens (such as GIMP) should present an easy, obvious way to navigate between screens.
- The filesystem is not as visible for a user as it is on a normal desktop environment. Consequently, users should *never* need to know a filename; they should only interact with metadata (thumbnails for the image viewer, song/artist name for music, from/to/subject headers in e-mail, and so on).
- Wherever possible, traditional open/save dialogs should be avoided — although (given the most popular iPhone user requests) the ability to copy and paste between applications is important.
- "Attach File" dialogs need to present a friendly list of objects to attach (thumbnails or metadata).
- If the user has to be presented with a list of files, the list must include only relevant files. Hidden files and folders should always be excluded.
- The number of configuration options should be minimized. Applications should come preconfigured with intelligent defaults. Options should be included only if they're easily understood by the computer-phobic or if they're likely to be useful to a large number of users.
- Users love instant feedback. Aim for any interaction with the UI results in a visual/audio feedback within 200ms, which is the upper limit of what's necessary for the appearance of "instant." For a tactile device like a MID, it's important that widgets act like physical objects. Delayed reactions remind the user that they're on a computer.
- Battery life is a major goal for the Ubuntu Mobile platform. Applications should use the application request latency from Device Kit, see Chapter 3.

Chapter 4: Application Development

- ❑ Error messages should suggest a course of action to the user in the event of a problem. The new Jaunty Notifications system is discussed later in this chapter.
- ❑ When interacting with a touchscreen, research has shown that users will prefer to use the pad of their finger rather than the very tip. This is important as it means interface elements should be no smaller than 1cm (0.4").

It is important to note that the minimum button size for a finger-driven UI is a function of the real pixel density, not the screen resolution. Pixel density is a measurement of the resolution of devices in various contexts and the program xdpyinfo can give a measure of this. Running the following on a MID machine

```
$ xdpyinfo | grep resolution
```

gives the following resolution used:

```
resolution:    96x96 dots per inch
```

Free Desktop Standards

Files need to be placed in specific locations within the file system so that new applications can be discovered, presented, and launched by the operating system. On Ubuntu, this placement of files conforms to freedesktop.org standards. These are a set of specifications for the interoperability of projects on X Window systems.

The most important standards to consider for a developer are:

The Desktop Entry Specification

This is a UTF-8 encoded file with the `.desktop` extension, which has a group header called `Desktop Entry` and a number of Entries in `Key=Value` format. The REQUIRED keys in this file are:

`Type` – Either an Application, Link or Directory

`Name` – The name of the application

`Exec` – The application to execute

(URL) – Is required if the `Type` (above) is set to Link

The `Exec` key must contain a command to execute, with either the full path to the executable itself or with just the name, if it can be found in the `$PATH` environment variable.

The `$PATH` *variable on Ubuntu by default is*

```
/usr/local/sbin:/usr/local/bin:/usr/sbin:/usr/bin:/sbin:/bin:/usr/games:
```

It is also possible to add special symbols on the command line, which will be expanded by the program launcher at the time of execution. For example, to pass a URL to a binary called pumd which is in the /usr/bin directory, the `Exec` key will look like:

```
Exec=pumd %u
```

Chapter 4: Application Development

The filename of the icon to be displayed, along with the name in the applications menu, if any (the `Icon` key), can also be included. Here is an example of this from the Hello-World .desktop file discussed later in this chapter:

```
[Desktop Entry]
Encoding=UTF-8
Name=Hello-World
Comment=Introduction to Hildon Applications
Exec=hello-world
Icon=hello.png
StartupNotify=true
Terminal=false
X-MultipleArgs=false
Type=Application
Categories=Application;Other
```

The Desktop Application Autostart Specification

Applications will be auto-started by a system if a .desktop file is placed in an autostart directory. The system wide setting for this (the variable `$XDG_CONFIG_DIRS`) is, by default, `/etc/xdg/autostart/`. The per user setting for this (the variable `$XDG_CONFIG_HOME`) is, by default, `~/.config/autostart/`.

Applications will be started after a user logs in.

XDG Base Directory Specification

This specifies where files should be located in relation to a base directory. This base directory for a user's data is specified in the environment variable `$XDG_DATA_HOME`, and for a user's configuration files in the environment variable `$XDG_CONFIG_HOME`.

The default value for `$XDG_DATA_HOME` is `$HOME/.local/share`, and for `$XDG_CONFIG_HOME` it is `$HOME/.config`.

Similarly, a set of preference-ordered base directories relative to which configuration files should be searched is defined by the environment variable `$XDG_CONFIG_DIRS`.

> This specification also defines directories relative to which data files should be searched, `$XDG_DATA_DIRS`, and relative to which non essential data should be cached, `$XDG_CACHE_HOME`.

Desktop Menu Specification

This can conceptually be thought of as an XML configuration file with the extension .menu, which defines the layout of menu items (.desktop files) and whether these items should be displayed in the menu. The XML file is located by default in

`$XDG_CONFIG_DIRS/menus/${XDG_MENU_PREFIX}applications.menu`

with the `$XDG_MENU_PREFIX` variable meaning that a user's menu can override (or rather be merged) with a system menu. The .desktop files are located in `$XDG_DATA_DIRS/applications/` by default.

This specification adds the fields `Categories`, `OnlyShowIn` and `NotShowIn` to the Desktop Entry Specification discussed above. `Categories` is a way to group menu items, `OnlyShowIn` is a way to make items only appear in specific environments, and `NotShowIn` the reverse.

Chapter 4: Application Development

In addition to knowing these specifications, it is also important to know what frameworks are available to use when developing an application.

Hildon: An Application Framework for Handheld Devices

The Hildon Application Framework is designed for small devices and has been widely used on this form factor. The framework had strong support from Nokia, is a part of GNOME Mobile, and has been extended to better support stylus-based usage and high display pixel density.

Basic GNOME technologies are used such as GTK as well as the GConf configuration system, and both of these are discussed later in the chapter.

Hildon then is an extended and modified GTK, which is referred to as the "Hildon widget set." It effectively provides a desktop environment for mobile devices.

> *On Hildon, one notable deviation from traditional desktop GNOME is the replacement of bonobo and its related technologies built around the usage of CORBA with D-Bus.*

What Is Hildon in Terms of Code?

A `HildonProgram` is a `Gobject` representing the whole application. In this context, it is important to note that a `HildonProgram` is *not* a `GtkWidget`.

Only one `HildonProgram` can be created per process. It is accessed with `hildon_program_get_instance`.

`HildonWindows` are registered to the `HildonProgram` using `hildon_program_add_window()`. As the method name suggests, several `HildonWindows` are likely to be registered to a `HildonProgram`.

The following C code highlights some key concepts and functionalities that are available when working with the framework.

Creating the Program

Here's the code:

```
/* Start the container and window */
HildonProgram *program;
HildonWindow *window;
/* Start up GTK */
gtk_init (&argc, &argv);
/* Make a Hildon program */
program = HILDON_PROGRAM (hildon_program_get_instance());
```

Chapter 4: Application Development

Menus

Menus are created externally using `GtkMenu`. They are then added to a `HildonWindow` with `hildon_window_set_menu()`.

An example of this looks like the following:

```
GtkMenu *menu;
menu = gtk_menu_new ();
fill_in_menu (menu);
hildon_window_set_menu (menu);
```

Toolbars

Toolbars are created using a call to `hildon_window_add_toolbar()`. Consider the following example:

```
GtkToolbar *toolbar;
toolbar = create_toolbar ();
/* Adding the toolbar */
hildon_window_add_toolbar (window, toolbar);
```

Window-Specific Settings

Window-specific commands can be sent to the main window manager and task navigator. Applications can tell the task navigator that some window requires the user's attention. This will trigger the blinking of the window's icon in the task navigator.

This is shown by the following:

```
gtk_window_set_urgency_hint (GTK_WINDOW (window), TRUE);
```

Similarly, applications are able to change the icon representing each `HildonWindow` in the task navigator by using the `gtk_window_set _icon` set of functions.

The following code demonstrates this:

```
GdkPixbuf *icon = create_icon();
gtk_window_set_icon (GTK_WINDOW (window), icon);
```

Also, the title can be set per window with the following:

```
gtk_window_set_title (GTK_WINDOW (window), "Window Title");
```

If the application name was set with `g_set_application name()`, it will precede the window title in the title bar.

Program-Wide Settings

The `HildonProgram` object provides the programmer with the capability to set program-wide settings for all the registered `HildonWindows`.

Chapter 4: Application Development

The programmer can create a common menu, which will be shared (memory and functionality-wise) by all the windows. A window can override this common menu by having its own menu. The following code demonstrates this:

```
GtkMenu *common_menu = create_common_menu ();
GtkMenu *window_specific_menu = create_window_specific_menu ();

hildon_program_set_common_menu (program, common_menu);
hildon_window_set_menu (second_window, window_specific_menu);
```

Also, a common toolbar can be shared among all `HildonWindow`s. Windows can also add window-specific toolbars. The common toolbar will be displayed on the bottom of the stack of toolbars. The following code demonstrates this:

```
GtkToolbar *common_toolbar = create_common_toolbar ();
GtkToolbar *window_specific_toolbar = create_window_specific_toolbar ();
hildon_program_set_common_toolbar (program, common_toolbar);
hildon_window_add_toolbar (second_window, window_specific_toolbar);
```

Hibernation

A programmer can tell the Hildon Task Navigator whether or not an application is ready to be set to hibernation (background-killed) by calling `set_can_hibernate()`.

Putting Hildon Together

The following program will build up a program from scratch to show the composition of an application that fits into the Hildon Framework.

Hello World

Hello World is, well ... a hello world program. It is a simple C program with an application icon and a .desktop file.

First, create the boilerplate Hildon code.

Source

The source code is:

```
#include <hildon/hildon-program.h>
#include <hildon/hildon-window.h>
#include <gtk/gtkmain.h>

int main(int argc, char *argv[])
{
HildonProgram *program;
HildonWindow *window;
GtkWidget *label;

gtk_init(&argc, &argv);

program = hildon_program_get_instance();
```

```
        window = HILDON_WINDOW(hildon_window_new());
        hildon_program_add_window(program, window);

        g_set_application_name("Hello World");

        label = gtk_label_new("Hello World");
        gtk_container_add(GTK_CONTAINER(window), label);
        gtk_widget_show_all(GTK_WIDGET(window));

        g_signal_connect(G_OBJECT(window), "delete_event", G_CALLBACK(gtk_main_quit),
        NULL);

        gtk_main();

        return 0;
        }
```

Save the code as hello-world.c and make sure you have build-essential, libgtk2.0-dev, and libhildon-1-dev installed. Then compile the binary by using the following:

```
# gcc -o hello-world hello-world.c $(pkg-config --cflags --libs gtk+-2.0 hildon-1)
```

After compilation, run hello-world.

Output

The application executes and presents the interface that's shown in Figure 4-2.

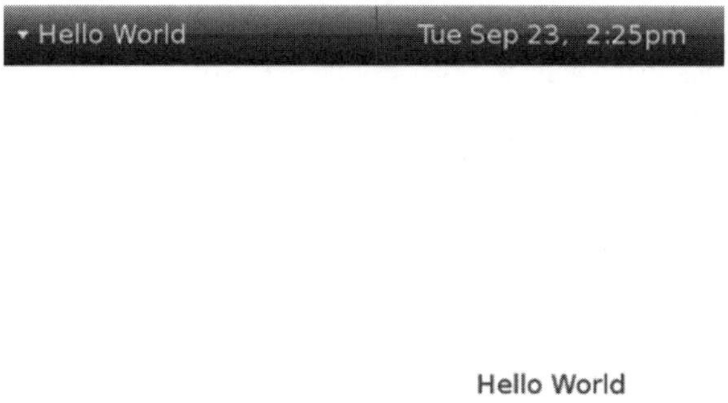

Hello World

Figure 4-2

Menu

Now create a standard GtkMenu and attach it to the HildonWindow. To do this, create an attach_menu function like this:

```
        static void attach_menu(HildonWindow *window)
        {
        GtkWidget *menu;
        GtkWidget *item_quit;
        if(NULL == window)
        return;
        menu = gtk_menu_new();
        item_quit = gtk_menu_item_new_with_label("Quit");
```

Chapter 4: Application Development

```
    gtk_menu_append(menu, item_quit);
    hildon_window_set_menu(window, GTK_MENU(menu));
    gtk_widget_show_all(GTK_WIDGET(menu));
}
```

Add this function into hello-world.c and call it in `main` before displaying the window:

```
attach_menu(window);
gtk_widget_show_all(GTK_WIDGET(window));
```

The menu looks like Figure 4-3.

Figure 4-3

Toolbar

Finally, create an `attach_toolbar` function to attach the toolbar to the `HildonWindow`:

```
static void attach_toolbar(HildonWindow *window)
{
GtkWidget *toolbar;
GtkToolItem *item_quit;
if(NULL == window)
return;
toolbar = gtk_toolbar_new();
item_quit = gtk_tool_button_new_from_stock(GTK_STOCK_QUIT);
gtk_toolbar_insert(GTK_TOOLBAR(toolbar), item_quit, -1);
hildon_window_add_toolbar(window, GTK_TOOLBAR(toolbar));
}
```

Add this to the source code and then call it after the `attach_menu()` function created earlier:

```
attach_menu(window);

attach_toolbar(window);
```

The toolbar looks like Figure 4-4.

Figure 4-4

Here's the final source code:

```c
#include <hildon/hildon-program.h>
#include <hildon/hildon-window.h>
#include <gtk/gtkmain.h>

static void attach_menu(HildonWindow *window)
{
GtkWidget *menu;
GtkWidget *item_quit;
if(NULL == window)
return;

menu = gtk_menu_new();
item_quit = gtk_menu_item_new_with_label("Quit");
gtk_menu_append(menu, item_quit);
hildon_window_set_menu(window, GTK_MENU(menu));
gtk_widget_show_all(GTK_WIDGET(menu));
}
static void attach_toolbar(HildonWindow *window)
{
GtkWidget *toolbar;
GtkToolItem *item_quit;
if(NULL == window)
return;
toolbar = gtk_toolbar_new();
item_quit = gtk_tool_button_new_from_stock(GTK_STOCK_QUIT);
gtk_toolbar_insert(GTK_TOOLBAR(toolbar), item_quit, -1);
hildon_window_add_toolbar(window, GTK_TOOLBAR(toolbar));
}

int main(int argc, char *argv[])
{
HildonProgram *program;
HildonWindow *window;
GtkWidget *label;
gtk_init(&argc, &argv);
program = hildon_program_get_instance();
window = HILDON_WINDOW(hildon_window_new());
hildon_program_add_window(program, window);
g_set_application_name("Hello World");
label = gtk_label_new("Hello World");
gtk_container_add(GTK_CONTAINER(window), label);
attach_menu(window);
attach_toolbar(window);
gtk_widget_show_all(GTK_WIDGET(window));
g_signal_connect(G_OBJECT(window), "delete_event", G_CALLBACK(gtk_main_quit),
NULL);
gtk_main();
return 0;
}
```

Run the compilation again:

```
# gcc -o hello-world hello-world.c $(pkg-config --cflags --libs gtk+-2.0 hildon-1)
```

Chapter 4: Application Development

Move the binary created to /usr/bin and add the .desktop file into /usr/share/applications. The .desktop file for the hello-world application looks like this:

```
[Desktop Entry]
Encoding=UTF-8
Name=Hello-World
Comment=Introduction to Hildon Applications
Exec=hello-world
Icon=hello.png
StartupNotify=true
Terminal=false
X-MultipleArgs=false
Type=Application
Categories=Application;Other
```

Create an icon 48 × 48 using GIMP and save it as hello.png. Place the icon for the application into /usr/share/icons/hicolor/<size>/<type> (for example, /usr/share/icons/hicolor/48x48/apps/hello).

The .desktop file and the icon must have read permissions. There must also not be another "hello" in your icons directory as there will be a conflict.

The code presents the icon on the desktop, as shown in Figure 4-5.

Figure 4-5

When the icon is clicked, it executes the application.

Other Toolkits

If the new application being written is for a device without a touchscreen or one that will run a distribution such as Ubuntu Netbook Remix, then it is possible to use pure GTK as a framework.

GTK+ is a toolkit designed to facilitate the development of applications using a standard graphics user interface. It was initially created as a tool used in the production of The GIMP, a graphics design package — GTK stands for "GIMP ToolKit."

It is useful for producing far more than just The GIMP, however, and has been applied to almost every kind of GUI application imaginable.

A GTK environment handles user requests by introducing an "event loop", which is a stack of the various input instructions the user provides, such as keystrokes and mouse clicks, and provides a set of handling functions that define what action to take when that instruction is received. When the program is run, the handling functions are set up and then the loop is started, and keeps running until a "quit" event is processed (for example, when the "Exit" entry in a menu is pulled down and selected).

Chapter 4: Application Development

This "event loop" is started by calling the function `gtk_main();` which starts up the event loop, the details of which are hidden from the user. GTK is an event-driven toolkit, which means it will sleep in `gtk_main();` until an event occurs and control is passed to the appropriate function. This passing of control is done using the idea of "signals."

Signals

Signals are processed by attaching a "callback" function to a given signal for any particular object (widget). For example, after creating a new button

```
button = gtk.Button()
```

a callback function that acts when the button is pressed can be set up by calling the following:

```
button.connect("clicked", self.callback, "a button")
```

which connects the "clicked" signal of the button to a callback.

The use of signals can also be seen in the "Putting All the Concepts Together" section later in this chapter. The first two lines of the code below connect to the status icon, and then signals are created which detect drive mounting. The final three lines connect to the signals already created.

```
self.connect('activate', self._on_activate_event)
self.connect('popup-menu', self._on_popup_menu_event)

# add a signal to detect when a cd is inserted into the drive
  gobject.signal_new('device-is-optical-disc', gtk.StatusIcon,
    gobject.SIGNAL_RUN_LAST, gobject.TYPE_NONE,
    (gobject.TYPE_STRING , gobject.TYPE_BOOLEAN))
# add a signal to detect when a device is mounted, so an action can be
# performed after that
gobject.signal_new('device-mounted', gtk.StatusIcon,
    gobject.SIGNAL_RUN_LAST, gobject.TYPE_NONE, (gobject.TYPE_STRING,))
gobject.signal_new('device-unmounted', gtk.StatusIcon,
    gobject.SIGNAL_RUN_LAST, gobject.TYPE_NONE, (gobject.TYPE_STRING,))

self.connect('device-is-optical-disc',
        self._on_optical_disc_insert_event)
self.connect('device-mounted', self._on_device_mounted_event)
self.connect('device-unmounted', self._on_device_unmounted_event)
```

Here are the method signatures

```
def _on_optical_disc_insert_event(self, status_icon, device, status):
   ...

def _on_device_mounted_event(self, status_icon, device):
   ...

def _on_device_unmounted_event(self, status_icon, device):
   ...
def _on_activate_event(self, status_icon):
   ...
def _on_popup_menu_event(self, status_icon, button, activate_time):
   ...
```

Chapter 4: Application Development

Layout

Rather than explicitly designing the layout of an application, defining the specific coordinates and dimensions of each element of the application, GTK+ uses relative positioning to indicate the placement of objects relative to each other. "Packing boxes" are the main objects that are used to achieve this placement.

These are invisible widget containers that we can pack our widgets into which come in two forms, a horizontal box, and a vertical box. You may use any combination of boxes inside or beside other boxes to create the desired effect.

Horizontal Boxes

When packing widgets into a horizontal box, the objects are inserted horizontally from left to right:

```
gtk_hbox_new()
```

A horizontal box uses the functions `gtk_box_pack_start()` and `gtk_box_pack()` to control layout and spacing.

Vertical Boxes

In a vertical box, widgets are packed from top to bottom:

```
gtk_vbox_new()
```

A vertical box uses the functions `gtk_box_pack_start()` and `gtk_box_pack()` to control layout and spacing.

Glade

In practice, laying out an application is often done using Glade. The Glade Interface Designer is a GNOME application that allows laying out widgets visually rather than through C code. Glade can generate C or C++ source code, which compiles to build the application; however, that method is no longer the best way to use Glade. Instead, the Glade project file, a .glade file in XML format, is parsed by the application at runtime using libglade. This allows you to modify/update the UI without recompiling the entire program and allows the program to be separate from the interface.

A Glade file is used in an application as follows:

```
/* create GladeXML object and connect signals */

    gxml = glade_xml_new (GLADE_FILE, NULL, NULL);

    glade_xml_signal_autoconnect (gxml);
```

The call to `glade_xml_new` creates a new `GladeXML` object, which `gxml` references, and it also creates all the widgets in the Glade file, which we pass to the function as the first argument. This would be declared somewhere in the application, as in the following:

```
/* this will hold the path to the glade file after the "make install" */

#define GLADE_FILE PACKAGE_DATA_DIR"/gnome3/gnome3.glade"
```

Chapter 4: Application Development

After importing the .glade file, you can access the widgets using the names you gave them in Glade and then use them in your program.

It's important to consider how to deal with the fact that Glade cannot create a Hildon Window when using the Hildon Framework. It can create only a GTK Window, and indeed it requires one as a parent for most other widgets (but not menus).

In a C application, this process would look like the following:

1. Manually modify the Glade file with window class from `GtkWindow` to `HildonWindow`.
2. In your C file, add `#include <glade/glade-build.h>`.
3. In your C file, add the following:

    ```
    static GtkWidget* glade_hildon_window_new(GladeXML *xml, GType type, GladeWidgetInfo *info) {
            return hildon_window_new();
    }
    ```

4. Before your window initialization, do the following:

    ```
    glade_register_widget(HILDON_TYPE_WINDOW,
                    glade_hildon_window_new,
                    glade_standard_build_children,
                    NULL);
    ```

5. Add the window to the Hildon program with

    `hildon_program_add_window()`

In Glade, you create the GTK Window and add widgets to it.

Glade can be used with Python as well as C; the process for using it with Hildon is the same — you "reparent" the GTK Window's child widgets to your real Hildon Window. To minimize reparenting, the following example program has only a "root" window with one "child," a GTK VBox widget, which in turn contains child widgets.

So, this VBox must be reparented from the GTK Window to a Hildon Window. The following Python example — called pumdGlade — shows how this is done.

Source

This is the source code for pumdGlade.

```
import pygtk
import gtk
import hildon
import gtk.glade

class pumdGlade():
    """pygtk-glade-hildon example for Professional Ubuntu Mobile Development book"""

    def __init__(self):
```

Chapter 4: Application Development

```python
            #make the hildon program
            self.program = hildon.Program()
            self.program.__init__()

            #make the hildon window and add to program
            self.window = hildon.Window()
            self.window.set_title("Professional Ubuntu Mobile Development")
            self.program.add_window(self.window)
            #receive signal to close window from framework close button

            if (self.window):
                self.window.connect("destroy", gtk.main_quit)

            #import the glade file and assign to self.wTree
            self.glade_file = "pygladeui.glade"
            self.wTree = gtk.glade.XML(self.glade_file)

            #reparent the vbox1 from glade to self.window
            self.vbox1 = self.wTree.get_widget("vbox1")
            self.reparent_loc(self.vbox1, self.window)

            #get menu from glade and reparent as common menu in hildon program
            self.menuGlade = self.wTree.get_widget("menu1")
            self.program.set_common_menu(self.menuGlade)

            #get quit menu item and connect signal
            self.menuItem_quit = self.wTree.get_widget("quit1")
            self.menuItem_quit.connect("activate", gtk.main_quit)

            #get hbox1 in order to modify contents based on user actions
            self.hbox1 = self.wTree.get_widget("hbox1")

            #get label1 for use
            self.label1 = self.wTree.get_widget("label1")

            #destroy the gtk window imported from glade
            self.gtkWindow = self.wTree.get_widget("window1")
            self.gtkWindow.destroy()

            #display everything
            self.window.show_all()

    def run(self):
        gtk.main()

# signal handlers
    def menuItem_quit1_pressed(self, widget):
        gtk.main_quit

#utility
    def reparent_loc(self, widget, newParent):
        widget.reparent(newParent)

if __name__ == "__main__":
    app = pumdGlade()
    app.run()
```

Chapter 4: Application Development

The program has one class, `pumdGlade`, whose constructor creates the `HildonProgram` and `HildonWindow`, imports the .glade file, assigns Glade objects to local names, reparents the widget, sets the Hildon program menu from a Glade-defined menu, and sets up signal handlers.

The class is instantiated as an object named `app`, and its `run()` method is executed.

Handling the .glade File

The .glade file is imported inside `pumdGlade`'s constructor. The code looks like this:

```
#import the glade file and assign to self.wTree
self.glade_file = "pygladeui.glade"
self.wTree = gtk.glade.XML(self.glade_file)
```

The first line defines a variable (`self.glade_file`) and assigns it the filename of the .glade file, in this case `pygladeui.glade`.

The second line actually imports all user interface objects from the specified .glade file into a new variable named `self.wTree`.

Now, all the objects defined in the .glade file can be retrieved from `self.wTree` and can be assigned local names for programmatic use.

The start of the Glade file looks like this:

```
<?xml version="1.0" standalone="no"?> <! — *- mode: xml -* — >
<!DOCTYPE glade-interface SYSTEM "http://glade.gnome.org/glade-2.0.dtd">
<glade-interface>
<widget class="GtkWindow" id="window1">
  <property name="visible">True</property>
  <property name="title" translatable="yes">window1</property>
  <property name="type">GTK_WINDOW_TOPLEVEL</property>
  <property name="window_position">GTK_WIN_POS_NONE</property>
  <property name="modal">False</property>
  <property name="resizable">True</property>
  <property name="destroy_with_parent">False</property>
  <property name="decorated">True</property>
  <property name="skip_taskbar_hint">False</property>
  <property name="skip_pager_hint">False</property>
  <property name="type_hint">GDK_WINDOW_TYPE_HINT_NORMAL</property>
  <property name="gravity">GDK_GRAVITY_NORTH_WEST</property>
  <property name="focus_on_map">True</property>
  <property name="urgency_hint">False</property>
  <child>
    <widget class="GtkVBox" id="vbox1">
      <property name="visible">True</property>
      <property name="homogeneous">False</property>
      <property name="spacing">0</property>
```

which shows the setting of some property names and the creation of the child widget.

Chapter 4: Application Development

The following Python code shows the re-parenting:

```
#reparent the vbox1 from glade to self.window

self.vbox1 = self.wTree.get_widget("vbox1")
self.reparent_loc(self.vbox1, self.window)
```

The first line gets the specified GTK `VBox` and assigns it to the `self.vbox1` variable. The second reparents the VBox from the GTK window made in Glade to the Hildon Window in Python called `self.window` with the `reparent_loc()` function defined here:

```
def reparent_loc(self, widget, newParent):

    widget.reparent(newParent)
```

The finished application when run on Ubuntu MID looks like Figure 4-6.

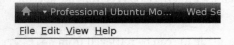

Professional Ubuntu Mobile Development

Figure 4-6

Another toolkit causing some excitement in mobile development communities is Clutter. This was originally developed by some GNOME developers at OpenedHand and is now a part of the Moblin project and developed at the Intel Open Source Technology Center.

Clutter

Clutter is a GObject-based library for creating animated hardware-accelerated graphical user interfaces. It relies upon OpenGL (1.4+) or OpenGL ES (1.1 or 2.0) for rendering, can be compiled on different platforms (Linux, Mac OS, and Win32) and has multiple bindings to other languages (including Python, Ruby, and Vala). It also supports media playback using Gstreamer and 2D graphics rendering using Cairo.

Clutter works by having a stage (a window) and then adding actors (widgets) to the stage and manipulating via the actor API. Actors can contain child actors (`ClutterGroup`, for example) and be manipulated as a whole.

Animations and visual effects can be created via the use of timelines and behaviors. Timelines provide accurate frame-based animations. Behaviors further extend this by taking a timeline and applying a control function (`clutter.Alpha`) to actors to modify some property of the actor in respect to time.

Chapter 4: Application Development

Clutter also has support for colors, signals, and labels, allowing for the creation of compelling visual interfaces without a developer, designer, or OEM needing to resort to proprietary and costly software such as Flash.

The following Python application shows the creation of three actors on a stage. This stage is embedded inside a GTK application. In the application, one image actor circles around the other in a permanent loop while the third (a text actor) fades in and out over time.

At the time of this writing, the latest stable version of Python Clutter is 0.8 and is either available at http://www.clutter-project.org/sources/pyclutter/0.8/ or installed using apt:

```
$ sudo apt-get install python-clutter
```

If it is installed from the source, use the following:

```
$ ./configure
$ make && sudo make install
```

Also make sure that the dependencies are installed:

```
sudo apt-get install python-cairo-dev libclutter-gst-0.8-dev libclutter-cairo-0.8-dev libclutter-0.8-dev python-gtk2-dev libclutter-gtk-0.8-dev
```

Next, create a Python file and import the required Clutter libraries:

```python
import sys
import cluttergtk    # must be the first to be imported
import clutter
import cluttercairo
import math
import array
import os, tempfile
import gtk
import cairo
```

The order of the import directives is relevant because of the required initialization process of the underlying C libraries. The correct order for importing the Clutter and related modules is:

```
first ↑
      | import cluttergtk
      | import cluttergst
      | import clutter
      | import cluttercairo
 last ↓
```

If you are importing the `gst` *module, you must import it after the* `cluttergst` *module.*

If you are importing the `gtk` *module, you must import it after the* `cluttergtk` *module.*

The main method first creates a GTK window and sets up a signal to quit the window. It then creates a packing box and packs the Clutter stage in it. Notice the call to `embed.realize()`.

Chapter 4: Application Development

To "activate" the widget before it can be used, do this:

```python
def main ():
    window = gtk.Window()
    window.connect('destroy', gtk.main_quit)
    window.set_title('Professional Ubuntu Mobile Development')
    vbox = gtk.VBox(False, 6)
    window.add(vbox)
    embed = cluttergtk.Embed()
    vbox.pack_start(embed, True, True, 0)
    embed.set_size_request(700, 600)

    # we need to realize the widget before we get the stage
    embed.realize()
    stage = embed.get_stage()
    stage.set_color(clutter.Color(228, 204, 152, 255))
```

The application then calls the class `UpstreamAndDownstream()`.

This class creates actors, sets up a timeline, applies behavior to an actor, and creates an actor on-the-fly from some text passed into the class:

```python
group = UpstreamAndDownstream()
```

The text on-the-fly is interesting to look at. A Clutter label is typically constructed and placed at the center of a stage like this:

```python
#create a clutter label
label = clutter.Label()
#set the labels font
label.set_font_name('Mono 32')
#add some text to the label
label.set_text("Mobile Linux")
#make the label brown
label.set_color(color_brown )
#put the label in the center of the stage
(label_width, label_height) = label.get_size()
label_x = (stage.get_width()/2) - label_width/2
label_y = (stage.get_height()/2) - label_height/2
label.set_position(label_x, label_y)
```

This application uses cairo, which is available through the library `cluttercairo` to draw a `ClutterTexture` directly onto an image. This image is saved to a temporary file and then imported as an actor into the scene.

It is called like this:

```python
create_text = group.render_title('Mobile Linux')
```

The `render_title` function looks like this:

```python
def render_title(self, text, size=80):
    # Make some variables
    size = int(size) * 3
```

Chapter 4: Application Development

```python
        font = "Meta"
        width,  height = 1500,   300
        def draw(cr):
            cr.set_source_rgba(1,1,1,0)
            cr.paint()
            # Some brown text
            cr.set_source_rgba(0.8,0.5,0,1)
            cr.select_font_face(font, cairo.FONT_SLANT_NORMAL,
    cairo.FONT_WEIGHT_NORMAL)
            cr.set_font_size(size)
            # We need to adjust by the text's offsets to center it.
            x_bearing, y_bearing, width, height = cr.text_extents(text)[:4]
            cr.move_to(-x_bearing,-y_bearing)
            # Technique to make the "thinner" parts more visible
            cr.text_path(text)
            cr.set_line_width(0.5)
            cr.stroke_preserve()
            cr.fill()
        return self.render_image(draw, width, height)
```

The preceding code positions the text and sets the size. It then calls the draw() function using some offsets to center it. Finally, it passes the draw to render_image(), which looks like this:

```python
    def render_image(self, drawer, width, height):
        # Eeck..cairo cannot stream..render it
        filename = tempfile.mkstemp()[1]
        # generic image render
        surface = cairo.ImageSurface(cairo.FORMAT_ARGB32, int(width), int(height))
        font_options = surface.get_font_options()
        font_options.set_antialias(cairo.ANTIALIAS_GRAY)
        context = cairo.Context(surface)
        # draw the context
        drawer(context)
        # Write the PNG data to our tempfile
        surface.write_to_png(filename)
        surface.finish()
        return filename
```

This type of functionality gives a user interface designer total control over fonts, and fonts are one area, particularly in embedded applications, that can really differentiate one application from another.

The class UpstreamAndDownstream() also in its __init__ method sets up a timeline, loops it, and sets up a couple of behaviors that can be used in the class:

```python
    def __init__ (self):
        clutter.Group.__init__(self)
        self.timeline = clutter.Timeline(100, 26)
        self.timeline.set_loop(True)
        self.alpha = clutter.Alpha(self.timeline, clutter.sine_func)
        self.moving = clutter.BehaviourEllipse(self.alpha, 200, 200, 400, 300, 0, 360)
        self.fade_opacity = clutter.BehaviourOpacity(opacity_start=0,
        opacity_end=255, alpha=self.alpha)
```

Chapter 4: Application Development

Back in the main method, some buttons are added to the vbox. The main window shows the loop that is started with `gtk.main()`.

The final animated application looks like Figure 4-7 on Ubuntu MID.

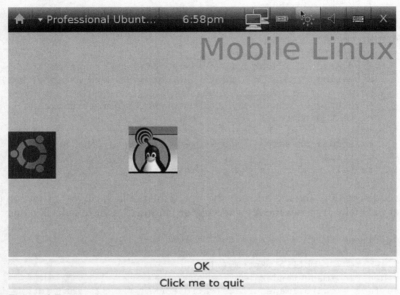

Figure 4-7

As this is a Python application, the .desktop file functions slightly differently. On the MID device, create a .desktop file:

```
[Desktop Entry]
Encoding=UTF-8
Name=Clutter
Comment= Professional Ubuntu Mobile Development
Exec=/usr/bin/clutterit
Icon=clutter.png
StartupNotify=true
Terminal=false
X-MultipleArgs=false
Type=Application
Categories=Application;Other
```

The file /usr/bin/clutterit is an executable script with

```
$ cd /home/ian/clutter
$ python gtk-clutter.py
```

Chapter 4: Application Development

Part of the appeal of Clutter is that it is cross-platform. The desire to "unite the desktops" is currently gaining some momentum. One toolkit which is cross-platform is QT, and the most visible use of QT is the KDE desktop.

> *GUADEC (Gnome) and Akademy (KDE) recently had a joint conference in Gran Canaria.*

The underlying technology then behind KDE is QT and Mark Shuttleworth has noted the following:

> *Qt will help us deliver ever more lustful applications to users. Nokia's continued investment in cross-platform Qt libraries, and the Linux platform, is a major driver of innovation in the free software desktop and mobile device stack.*

QT

Nokia recently acquired Trolltech and changed the license of QT from GPL v2 to the LGPL v 2.1 license.

OEMs can now link directly to QT libraries and not have all their linked code become automatically available under the GPL. They now have the option to pick and choose which code they want to share.

The following example was written in QT4 on Ubuntu Jaunty and is based on the tutorial at `http://wiki.python.org/moin/JonathanGardnerPyQtTutorial` but updated for QT4. It will

- ❑ Use QT Designer to generate Qt UI files.
- ❑ Use pyuic to generate Python programs.
- ❑ Use Qt Signals in Python.
- ❑ Create a simple application to interface a system command.

In order to start developing (in Python) with QT, make sure Python qt-dev, qt4.designer, and pyqt-tools are installed (`apt-get install python-qt-dev qt4.designer pyqt4.dev-tools`). Then start designer using:

```
$designer
```

This command will present Qt designer. This is somewhat similar to the Glade interface that you saw earlier, and functions in basically the same way. Drag UI elements onto the widget and right-click to change properties.

Create a new widget. Name it `pumd` and then add the following:

- ❑ **QLineEdit** — Name it **command** in the property dialog.
- ❑ **QPushButton** — Name it **schedule** in the property dialog. Change its text to "Schedule".
- ❑ **QDateTimeEdit** — Name it **time** in the property dialog.

When this is completed, save the file as pumd.ui in a new folder. The interface should look like Figure 4-8.

Chapter 4: Application Development

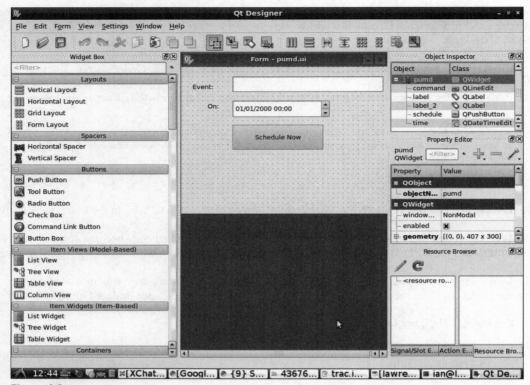

Figure 4-8

Then, in a shell, move into the folder and run

```
$ pyuic4 pumd.ui
```

This command will store the generated Python code that comes from the Qt UI file into pumd.py using

```
$ pyuic pumd.ui -o pumd.py
```

The file looks like this:

```
# -*- coding: utf-8 -*-
# Form implementation generated from reading ui file 'pumd.ui'
# Created: Sun Mar 22 16:11:48 2009
#      by: PyQt4 UI code generator 4.4.4
#
# WARNING! All changes made in this file will be lost!
from PyQt4 import QtCore, QtGui

class Ui_pumd(object):
```

Chapter 4: Application Development

```python
def setupUi(self, pumd):
    pumd.setObjectName("pumd")
    pumd.resize(407, 300)
    self.time = QtGui.QDateTimeEdit(pumd)
    self.time.setGeometry(QtCore.QRect(100, 70, 194, 34))
    self.time.setObjectName("time")
    self.command = QtGui.QLineEdit(pumd)
    self.command.setGeometry(QtCore.QRect(100, 20, 301, 32))
    self.command.setObjectName("command")
    self.schedule = QtGui.QPushButton(pumd)
    self.schedule.setGeometry(QtCore.QRect(100, 120, 181, 51))
    self.schedule.setObjectName("schedule")
    self.label = QtGui.QLabel(pumd)
    self.label.setGeometry(QtCore.QRect(20, 30, 71, 24))
    self.label.setObjectName("label")
    self.label_2 = QtGui.QLabel(pumd)
    self.label_2.setGeometry(QtCore.QRect(50, 70, 51, 24))
    self.label_2.setObjectName("label_2")
    self.retranslateUi(pumd)

    QtCore.QMetaObject.connectSlotsByName(pumd)

def retranslateUi(self, pumd):
    pumd.setWindowTitle(QtGui.QApplication.translate("pumd", "Form", None,
    QtGui.QApplication.UnicodeUTF8))
    self.schedule.setToolTip(QtGui.QApplication.translate("pumd", "Send Date",
    None, QtGui.QApplication.UnicodeUTF8))
    self.schedule.setText(QtGui.QApplication.translate("pumd", "Schedule Now",
    None, QtGui.QApplication.UnicodeUTF8))
    self.label.setText(QtGui.QApplication.translate("pumd", "Event:", None,
    QtGui.QApplication.UnicodeUTF8))
    self.label_2.setText(QtGui.QApplication.translate("pumd", "On:", None,
    QtGui.QApplication.UnicodeUTF8))
```

This is the GUI layout in Python code!

To make the development process easier, the previous command can be added into a `Makefile`; if changes are made in the UI, you only need to run `make` and the pumd.py will be automatically regenerated. The `Makefile` looks like the following code snippet (remember that inserting a <Tab> is important for the second line):

```
pumd.py: pumd.ui
    pyuic4 pumd.ui -o pumd.py
```

Now run it using the following:

```
$ make
```

Chapter 4: Application Development

Notice that it says something about all the files being up-to-date. Touch pumd.ui so it appears newer than pumd.py, and then run `make` again.

```
$ touch pumd.ui
$ make
```

With this in place, a GUI developer should be able to go and make changes to the GUI interface (such as moving things around) without affecting the logic behind the GUI. This makes for a clean separation of roles in development teams as all the designer needs to do is use Qt Designer to change the pumd.ui file, and run `make` to see the changes take effect.

So now the pumd.ui file and the pumd.py file have been created. It is now necessary to actually run the app. To do this, create the file pumd_at.py. This file looks like the following:

```python
from PyQt4 import QtCore, QtGui, Qt
from pumd import Ui_pumd
import sys, os
from time import localtime

class PumdAT(QtGui.QMainWindow):
    def __init__(self, parent=None):
        QtGui.QWidget.__init__(self, parent)
        self.ui = Ui_pumd()
        self.ui.setupUi(self)

        # Set the date to now
        now = Qt.QDateTime.currentDateTime()
        self.ui.time.setDateTime(now)

        # here we connect our signal on the 'schedule' button to the function
        #'schedule_clicked()'

        # which writes to the 'command'
      QtCore.QObject.connect(self.ui.schedule,QtCore.SIGNAL("clicked()"),
      self. schedule_clicked)

    def schedule_clicked(self):
        t = str(self.ui.time.dateTime().toString('hh:mm MM/dd/yyyy'))
        p = os.popen('at -m "%s"'%t, 'w')
        self.ui.command.setText('at -m "%s"'%t)
        if p:
            Qt.QMessageBox.information(self,
                "Event Scheduled", "Your event was scheduled with success",
                Qt.QMessageBox.Ok)
            return

if __name__ == "__main__":
    app = QtGui.QApplication(sys.argv)
    myapp = PumdAT()
    myapp.show()
    sys.exit(app.exec_())
```

Chapter 4: Application Development

When this file is run, it executes to present the interface that's shown in Figure 4-9. It allows a user to select a system command to be run.

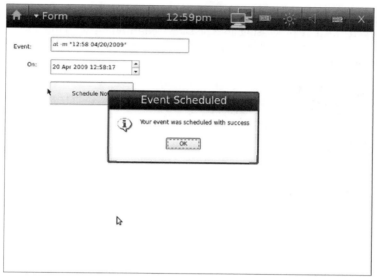

Figure 4-9

Another toolkit available for a mobile developer to use is the Enlightenment Foundation Libraries (EFL) or simply "e." These libraries form the basis of the Enlightenment Window Manager.

EFL

EFL is a toolkit that is well-suited to embedded development, as it has very low memory overheads. According to the developers, "Enlightenment provides the building blocks for creating beautiful applications."

EFL consists of:

- Evas
- Ecore
- Edje
- Embryo
- Eet

The EFL libraries fit well into the embedded device context and are particularly useful for animations, transparent components, and themeable applications. A user interacts with the window manager, which itself shows a compiled Edje theme. Events are captured by Ecore while Evas communicates directly with the kernel to efficiently control hardware resources such as 3D acceleration.

Chapter 4: Application Development

Themeable EFL applications are made using the Edje Layout engine, which sits above Evas, the X11 canvas library. Edje features a transparent, signal-based separation of interface and application logic so that every application that uses it is "skinnable."

An Edje interface file is just a single file arranged to accommodate interface information, images, and fonts. This file is generated by processing an Edje Data Collection file with the Edje Compiler. This makes theming a device very easy. An example of how a theme is laid out can be seen in the `main.edc` file for a window border:

```
#define COLOR_FG 246 151 67 255
#define COLOR_BG 245 245 245 255

fonts {
/* use the default */
}

collections {
#include "border.edc"
}
```

This includes another file called `border.edc`, which looks like this:

```
group {
    name, "e/widgets/border/default/border";

    parts {
        part {
            name, "main";
            type, RECT;
            description {
                state, "default" 0.0;
                color, COLOR_FG;
            }
        }
        part {
            name, "main_inner";
            type, RECT;
            description {
                state, "default" 0.0;
                color, COLOR_BG;
                rel1.offset, 1 1;
                rel2.offset, -2 -2;
            }
        }
    }
```

Note that this file is saved in the same directory as main.edc.

Here, the `main_inner` part hides all of `main`, except for the 1-pixel border. A key concept is the notion of "parts," which come together to form the theme.

A theme then can be compiled using:

```
$ edje_cc main.edc
```

Chapter 4: Application Development

The resulting .edj file is then copied to the "themes" directory — usually /usr/share/themes.

There is a web service for Canola (see below) which creates and then packages EFL themes for devices. It is available at `http://thememaker.openbossa.org`.

Canola

One of the best-known applications that uses the "e" framework is a media center–like application called Canola. This started life on the Nokia Internet Tablets where it has constantly been the most popular application among users. It was recently released under the GPL and has now been packaged for Ubuntu.

To run Canola, add the following:

```
deb http://ppa.launchpad.net/canola/ppa/ubuntu jaunty main
deb-src http://ppa.launchpad.net/canola/ppa/ubuntu jaunty main
```

to /etc/apt/source.list:

```
$ sudo apt-get update
$ sudo apt-get install canola2
```

For karmic and beyond, "upstream" for the Ubuntu MID project will be Mer and no longer Moblin. There is talk of including Canola as the default media player in Mer. Mer is built on a thin base of Ubuntu combined with the best open-source elements of Nokia's Maemo platform.

Elementary

A widget set which is part of the Enlightenment suite is Elementary. This is well suited for embedded devices as it has a small footprint and, although it is a small API, it is one which can be entirely themed using the power of Edje. This makes it very attractive for device manufacturers looking to highly customize a device.

Packages for Ubuntu are available from `http://packages.enlightenment.org`.

Add

```
deb http://packages.enlightenment.org/ubuntu jaunty main extras
```

to your /etc/apt/sources.list and then install elementary:

```
$ sudo apt-get install libelm
```

To get an idea about what can be done with this toolkit, run the test application

```
$ elementary_test
```

which displays the interface seen in Figure 4-10.

Chapter 4: Application Development

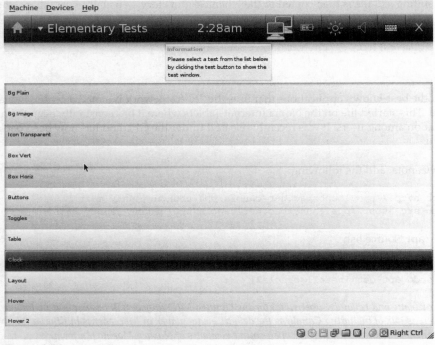

Figure 4-10

What Key Technologies Do I Need to Know to Develop Applications for a Mobile Device?

It is necessary to have a good understanding of some fundamental concepts in order to ensure the correct configuration and smooth functioning of an application on Ubuntu Mobile devices. The first of these is D-Bus.

D-Bus

D-Bus is an inter-process communication mechanism — a medium for local communication between processes running on the same host. D-Bus is meant to be fast and lightweight, and is designed for use as a unified middleware layer underneath the main free desktop environments.

There are two Bus types:

- **SystemBus** — This is a Bus service to connect daemons that offer system services. These services, generally, start when the machine boots.
- **SessionBus** — This is a Bus service to connect applications that are generally started by the user. Rhythmbox is an example of such an application.

Chapter 4: Application Development

Object Paths and Bus Names

When communicating over a bus, applications obtain a "service name": This is how the application chooses to be known by other applications on the same bus. The service names are brokered by the D-Bus bus daemon and are used to route messages from one application to another.

Because it is a binary protocol, D-Bus messages incur low overhead when marshaling data. Messages consist of two sections, the header and the body. The header contains the metadata for the message. This can include routing information and the type signature for the data. The body contains the data being sent.

Complex data types such as arrays and dictionaries can be encoded into a message. This allows communication between applications written in a wide range of languages with the ability to retain a common set of data structures.

Messages are sent to objects, not applications. Applications themselves are free to register as many objects as they wish. A D-Bus object can be thought of like any other object in a programming language with the exception that they are pointed to not by memory addresses but by object paths. Object paths take the form of a string that looks similar to UNIX filesystem paths.

They are slash-separated labels, each consisting of letters, digits, and the underscore character ("_"). They must always start with a slash and must not end with one. Here's an example:

```
/com/pumd/SimpleTextEditor
```

D-Bus objects are invoked through both methods and signals. D-Bus methods are similarly like any other method in an object-oriented language.

You invoke a method on an object directly by sending a method message to an object's interface on a service. The message may contain a list of parameters you wish to send to the method. A method can reply back both synchronously, where your program waits for a reply, or asynchronously where your program will be notified when a reply has been received. Methods do not have to send a reply back. Signals are simple notifications that an event has occurred. A signal is broadcast over the bus and therefore does not require a service to send to. Anyone listening for a particular signal will be notified when it is emitted.

An example which get properties and methods from DeviceKit D-Bus can be seen in the code from the "Putting All the Concepts Together" section:

```
def _get_dev_iface(self, object_path, interface):
    bus = dbus.SystemBus()
    proxy_obj = bus.get_object('org.freedesktop.DeviceKit.Disks',
            object_path)
    iface_obj = dbus.Interface(proxy_obj, interface)

    return iface_obj

def _get_dev_prop(self, device, property):
    props_iface_obj = self._get_dev_iface(device,
            'org.freedesktop.DBus.Properties')

    try:
```

Chapter 4: Application Development

```
            res = props_iface_obj.Get('org.freedesktop.DeviceKit.Disks.Device',
                    property)
        except dbus.DBusException:
            res = ''

        return res

    def _get_dev_methods(self, device):
        device_iface_obj = self._get_dev_iface(device,
                'org.freedesktop.DeviceKit.Disks.Device')

        return device_iface_obj
```

Exporting Objects with D-Bus

In order to export objects (to make them available to other applications), it is necessary that an event loop be running and that the D-Bus be connected to it. The following code from pydistcc shows a gobject main loop and a decorator which exports the method. Also notice the call to dbus.service.Object, which exports the method onto the Bus.

```
class DbusObject(threading.Thread, dbus.service.Object):
    def __init__(self, serverlist, condition):
        self.serverlist = serverlist
        self.condition = condition
        self.servicename = 'net.sourceforge.PyDistcc'
        self.objectpath = '/net/sourceforge/PyDistcc/DbusObject'

        threading.Thread.__init__(self)
    @dbus.service.method('net.sourceforge.PyDistcc', in_signature='',
                    out_signature='a{ss}')
    def currentserverlist(self):
        return self.serverlist

    def run(self):
        loop = gobject.MainLoop()
        self.bus = dbus.SystemBus()
        self.busname = dbus.service.BusName(self.servicename, bus=self.bus)
        dbus.service.Object.__init__(self, self.busname, self.objectpath)

        loop.run()
```

Chapter 4: Application Development

It is also possible to use a decorator to export a signal like:

```
@dbus.service.signal(dbus_interface='net.sourceforge.PyDistcc',signature='')
```

Connect to a D-Bus Signal

A default signal receiver looks like

```
bus.add_signal_receiver(self.on_device_added,
                        'DeviceAdded',
                        'org.freedesktop.DeviceKit.Disks',
                        'org.freedesktop.DeviceKit.Disks',
                        '/org/freedesktop/DeviceKit/Disks')
```

and below is a custom receiver. Notice the extra_parameters argument which allows extra information, which will be passed to the callback method.

```
bus.add_signal_receiver(
        lambda *args:self._on_device_changed(extra_parameters, *args),
        'DeviceChanged',
        'org.freedesktop.DeviceKit.Disks',
        'org.freedesktop.DeviceKit.Disks',
        '/org/freedesktop/DeviceKit/Disks',
        extra_parameters)
```

Useful D-Bus Command-Line Applications

There are some useful applications available in Ubuntu when working with DBUS.

D-Bus Viewer

D-Bus Viewer is a tool that lets you inspect D-Bus objects and messages. You can choose between the system bus and the session bus.

The application is available at http://launchpadlibrarian.net/6844344/dbus-viewer_4.2.3-0ubuntu3_i386.deb. Install it and then run the following:

```
$ dbus-viewer
```

There is a clone of `dbus-viewer` *with a GTK+ interface called D-Bus Explorer.*

This displays the D-Bus services, as shown in Figure 4-11.

Chapter 4: Application Development

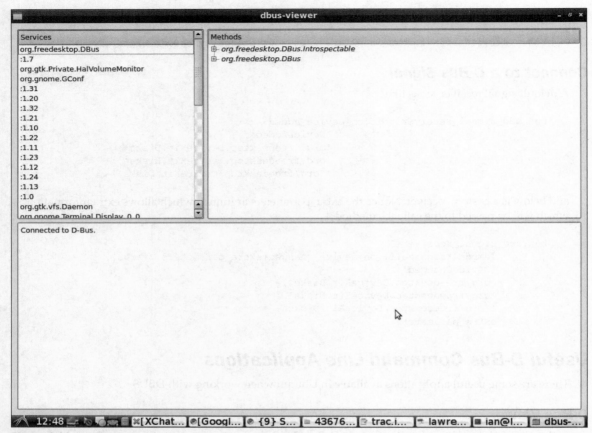

Figure 4-11

In the python-dbus-doc package, there is also a script called list-system-services.py that does what the filename suggests. This script is found in /usr/share/doc/python-dbus-doc/examples. Running it results in output like this:

```
$ python list-system-services.py
com.redhat.NewPrinterNotification
com.ubuntu.SystemService
fi.epitest.hostap.WPASupplicant
org.bluez
org.freedesktop.Avahi
org.freedesktop.ConsoleKit
org.freedesktop.DBus
org.freedesktop.Hal
org.freedesktop.NetworkManager
org.freedesktop.NetworkManagerSystemSettings
org.freedesktop.NetworkManagerUserSettings
org.freedesktop.SystemToolsBackends
org.x.config.display0
```

which are the current running D-Bus services.

Chapter 4: Application Development

D-Bus Send

D-Bus Send is used to send a message to a D-Bus message bus. Nearly all uses must provide the `--dest` argument, which is the name of a connection on the bus to send the message to. It can be used, for example, on the command-line to send Rhythmbox to the next track, as follows:

```
dbus-send --dest='org.gnome.Rhythmbox' /org/gnome/Rhythmbox/Player org.gnome.Rhythmbox.Player.next
```

D-Bus Monitor

The `dbus-monitor` command is used to monitor messages going through a D-Bus message bus. It has two different output modes, the "classic-style" monitoring mode and profiling mode. The profiling format is a compact format with a single line per message and microsecond-resolution timing information:

```
$ dbus-monitor -profile
```

It results in the following output when the power cord on a device is disconnected:

```
sig       1241300183      50361       2           /org/freedesktop/DBus       org.freedesktop
.DBus           NameAcquired
mc        1241300183      50599       347         :1.57       /org/freedesktop/indicate
org.freedesktop.DBus.Properties     Get
mc        1241300183      50634       5           :1.104      /org/freedesktop/DBus
org.freedesktop.DBus                AddMatch
sig       1241300186      367806      53          /org/freedesktop/PowerManagement
org.freedesktop.PowerManagement     OnBatteryChanged
sig       1241300186      369401      54          /org/freedesktop/PowerManagement
org.freedesktop.PowerManagement     PowerSaveStatusChanged
mc        1241300186      375244      55          :1.32       /org/gnome/ScreenSaver
org.gnome.ScreenSaver               Throttle
mr        1241300186      375301      31          55          :1.32
mc        1241300186      379997      56          :1.32       /org/gnome/ScreenSaver
org.gnome.ScreenSaver               SimulateUserActivity
mc        1241300186      385501      57          :1.32       /org/freedesktop/
Notifications   org.freedesktop.Notifications       GetCapabilities
mr        1241300186      389415      47          57          :1.32
mc        1241300186      389463      58          :1.32       /org/freedesktop/
Notifications   org.freedesktop.Notifications       Notify
mr        1241300186      480567      48          58          :1.32
sig       1241300186      697351      59          /org/freedesktop/PowerManagement/
Backlight       org.freedesktop.PowerManagement.Backlight       BrightnessChanged
sig       1241300188      728445      49          /org/freedesktop/Notifications
org.freedesktop.Notifications       NotificationClosed
```

Chapter 4: Application Development

D-Bus Launch

The `dbus-launch` command is used to start a session bus instance of dbus-daemon from a shell script (normally on a user login to a system). It is used like this:

```
## test for an existing bus daemon
        if test -z "$DBUS_SESSION_BUS_ADDRESS"; then
            ## if not found, launch a new one
            eval 'dbus-launch --sh-syntax --exit-with-session'
            echo "D-Bus per-session daemon address is: $DBUS_SESSION_BUS_ADDRESS"
        fi
```

D-Feet

D-feet (`sudo apt-get install dfeet`) is a D-Bus debugger written in PyGTK. It provides a GUI to connect to the system or session bus and the ability to execute methods with parameters on the bus and then see the return values. Figure 4-12 shows the effect of executing the `GetOnBattery()` method on the org.freedesktop.Power Management Session Bus when the device is running on battery power.

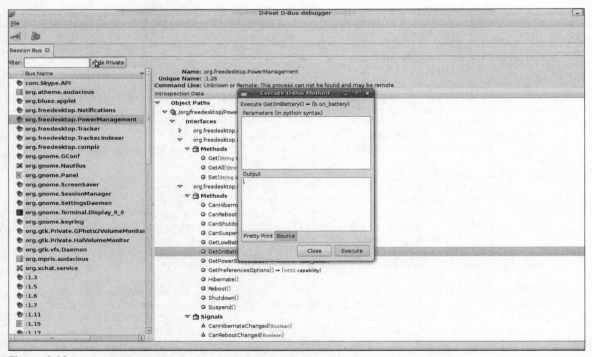

Figure 4-12

> *d_feet is highly recommended for debugging DBUS problems.*

Chapter 4: Application Development

D-Bus Security

The bus daemon has multiple instances on a typical computer. The system daemon has heavy security restrictions on what messages it will accept, and is used for system-wide communication. The other instances are created one per user login session. These instances allow applications in the user's session to communicate with one another.

Security is configured for these instances in a configuration file. This file lives in /etc/dbus-1/system.d.

To configure policies, create a file in this directory called application.conf with the policies (this file could be called anything — the name does not matter). This file will look like:

```
<!DOCTYPE busconfig PUBLIC
  "-//freedesktop//DTD D-BUS Bus Configuration 1.0//EN"
  "http://www.freedesktop.org/standards/dbus/1.0/busconfig.dtd">
<busconfig>
        <policy user="root">
                <allow own="net.sourceforge.PyDistcc"
                  send_destination="net.sourceforge.PyDistcc"
                  send_interface="net.sourceforge.PyDistcc"/>
        </policy>
        <policy user="rodrigo">
                <allow own="net.sourceforge.PyDistcc"
                  send_destination="net.sourceforge.PyDistcc"
                  send_interface="net.sourceforge.PyDistcc"/>
        </policy>
        <policy at_console="true">
                <allow send_destination="net.sourceforge.PyDistcc"
                  send_interface="net.sourceforge.PyDistcc"/>
        </policy>
        <policy context="default">
                <deny own="net.sourceforge.PyDistcc"
                  send_destination="net.sourceforge.PyDistcc"
                  send_interface="net.sourceforge.PyDistcc"/>
        </policy>
</busconfig>
```

The code example shows the use of policies. The default policy of the D-Bus system bus is as follows:

- ❏ Name ownership is DENIED by default.
- ❏ Method calls are DENIED by default.
- ❏ Replies to method calls, including errors, are PERMITTED by default.
- ❏ Signals are PERMITTED by default.

Therefore each service *must*, in its policy configuration, permit an appropriate user to own the name it wishes to claim — for example:

```
<policy user="rodrigo">
  <allow own="net.sourceforge.PyDistcc"/>
</policy>
```

Chapter 4: Application Development

It must also allow method calls to be made on objects it exports, for particular users. This can be done by allowing all method calls to the claimed name:

```
<policy user="rodrigo">
  <allow own="net.sourceforge.PyDistcc"
  send_destination="net.sourceforge.PyDistcc"
  send_interface="net.sourceforge.PyDistcc"/>
</policy>
```

It is important that `send_destination` be included on *all* allow or deny tags and omitting it is a potential security hole. Do not do this:

```
<allow send_interface="x.y.z" />
```

It would allow any service to receive method calls of the given interface!

PolicyKit

If the D-Bus service needs privileges (something requiring sudo) then it is possible to use PolicyKit to control this access.

PolicyKit (http://hal.freedesktop.org/docs/PolicyKit/PolicyKit.8.html) comes by default with Ubuntu and it is divided into two parts – the first part is a 'mechanism' (which runs as privileged) and the second part is the 'policy agent' (which runs as unprivileged) with the two parts interacting on the system bus. The 'mechanism' uses PolicyKit to determine whether to allow access to the process or the user which requested it. As an example the devicekit power QoS policy (which is defined in the XML file /usr/share/PolicyKit/policy/org.freedesktop.devicekit.power.qos.policy) has an 'action' concerning minimal latency

```
<action id="org.freedesktop.devicekit.power.qos.set-minimum-latency">
  <description>Set administrator settings for latency control</description>
  <message>Authentication is required to set administrator settings for latency
</message>
  <defaults>
    <allow_inactive>no</allow_inactive>
    <allow_active>auth_admin</allow_active>
  </defaults>
</action>
```

sending a D-Bus message like

```
$ dbus-send --session --print-reply ↵
            --dest=org.freedesktop.PolicyKit.AuthenticationAgent / ↵
            org.freedesktop.PolicyKit.AuthenticationAgent.ObtainAuthorization
            string:org.freedesktop.devicekit.power.qos.set-minimum-latency ↵
            uint32:0 ↵
            uint32:$PPID
```

pops up an interface asking for authorization. Changing

```
<allow_active>auth_admin</allow_active>
```

to

```
<allow_active>yes</allow_active>
```

returns the method without asking for authorization.

The allow_active tag specifies the default answer that PolicyKit will return for an active session.

The next key technology that is necessary to understand when developing applications for a mobile device is GConf.

GConf

GConf is a way to store configuration settings that ships with GNOME. Conceptually it resembles the Windows Registry, but GConf abstracts the registry concept into a library. This provides a simple interface to applications, and an architecture that tries to make things easy for OEMs, system administrators, and developers.

Essentially, GConf provides a preferences database, which is like a simple filesystem. The filesystem contains *keys* organized into a hierarchy. Each key is either a directory containing more keys, or has a value. For example, key /apps/metacity/general/titlebar_font contains an integer value that gives the size of the title bar font for the Metacity window manager.

The following Python application shows the use of GConf. It is an application that shows the most recent files used by a user, which it pulls from the GConf settings. The gconf-example.glade file is not shown here, but it is included in the code download on the Wrox site.

```python
#!/usr/bin/env python
# -*- coding: utf-8 -*-
import gtk
import gtk.glade
import gconf

class GconfApp(object):
    def __init__(self):
        #setting the glade file
        self.gladefile = 'gconf-example.glade'
        self.gconfwinfo = '/apps/gconf-example/winfo'
        self.gconfprefs = '/apps/gconf-example/prefs'
        self.wTree = gtk.glade.XML(self.gladefile, 'window1')

        signals = {
                'on_window1_delete_event': self.on_window1_delete_event,
                #'on_window1_delete_event': self.on_window1_destroy,
                'on_window1_destroy': self.on_window1_destroy,
            }

        self.wTree.signal_autoconnect(signals)

        self.window = self.wTree.get_widget('window1')
```

Chapter 4: Application Development

```python
            cl = gconf.client_get_default()
            width = cl.get_int(self.gconfwinfo + '/window_width') or 640
            height = cl.get_int(self.gconfwinfo + '/window_height') or 480
            self.window.set_default_size(width, height)

            self._hpaned = self.wTree.get_widget('hpaned1')
            if cl.get(self.gconfwinfo + '/' + self._hpaned.name):
                paned_position = cl.get_int(self.gconfwinfo + '/' + self._hpaned.name)
                self._hpaned.set_position(paned_position)
                self._hpaned.set_data('last_position', self._hpaned.get_position())

            self.window.show_all()

       def saveState(self):
            '''
            Save the dimensions of the main window, and the position of the panes.
            '''
            cl = gconf.client_get_default()
            cl.set_int(self.gconfwinfo + '/window_width', self.window.allocation.width)
            cl.set_int(self.gconfwinfo + '/window_height', self.window.allocation.height)
            cl.set_int(self.gconfwinfo + '/%s' % self._hpaned.name, self._hpaned.get_
position())

       def on_window1_delete_event(self, *args):
            self.saveState()

       def on_window1_destroy(self, *args):
            self.window.destroy()
            gtk.main_quit()

if __name__ == '__main__':

    gconfapp = GconfApp()
    gtk.main()
```

The preceding code functions in a client-server way. After first setting some signals, it gets the default GConf client. The actual values for the recent files that are shown in the UI are obtained from a Glade file with the call to

```xml
    <widget class="GtkRecentChooserWidget" id="recentchooser1">
                    <property name="visible">True</property>
                    <property name="limit">50</property>
    </widget>
```

The main advantage to using GConf is that the configuration data is stored in a tree structure we used previously, which makes it a lot easier to manage the preferences of a large application, with perhaps many preferences and configuration options. It also enables notifications across applications. This makes it possible for multiple applications, or multiple running instances of the same application, to immediately react to changes to preferences, and perform the necessary work to adapt to the new preferences.

Chapter 4: Application Development

The capability to react to changes and to show system changes in a coherent manner to a user was substantially improved during the Jaunty and karmic cycles through the use of a new notifications framework. This is the next key technology to understand.

Notifications

A new way to present information to a user was debuted in the Jaunty release of Ubuntu. This is a way to generate passive pop-ups (also called "poptarts"), which notify the user of some event.

While some information requires a response from the user (a low battery for instance), a large amount of the information generated by a system is purely informative and requires no user interaction. Notifications (provided through the notify-osd daemon or through the `org.freedesktop.Notifications` Dbus interface) provide for such a use case in that they do not provide "actions" for the user (they cannot be clicked and appear to hover and then fade). Semantically, the bubbles are either confirmations (brightness increased, for example) or notifications (new track playing).

Commands to the notification system can be sent manually using

```
$ notify-send
```

and also programatically using the D-Bus interface. There are four ways to manually use the `notify-send` tool (which is provided by the package `libnotify-bin`):

Icon - Summary - Body

```
$ notify-send 'Hello Rodrigo' -i notification-message-IM
```

Icon - Summary

```
$ notify-send 'WiFi connection lost' -i notification-network-wireless-disconnected
```

Summary - Body

```
$ notify-send 'PUMD' 'This is some notification'
```

Summary - only

```
$ notify-send 'Summary-only'
```

To send a notification programmatically (this is from the "Putting All the Concepts Together" section later in this chapter), use:

```
def __init__(self):
    (...)
    # saving notification system availablity status
    self._notification_available = pynotify.init('disk-applet')
    (...)

def on_device_added(self, device):
```

93

Chapter 4: Application Development

```
        if not self._get_dev_prop(device, 'device-is-drive') \
                and not self._notification_available:
            return

        if self._get_dev_prop(device, 'drive-connection-interface') == 'usb':
            notification_icon = 'notification-device-usb'
        elif self._get_dev_prop(device,
                'drive-connection-interface') == 'firewire':
            notification_icon = 'notification-device-firewire'
        else:
            notification_icon = 'notification-device'

        body_msg = '%s %s %s' % (
                self._get_dev_prop(device, 'drive-vendor'),
                self._get_dev_prop(device, 'drive-model'),
                self._get_dev_prop(device, 'drive-revision'))

        n = pynotify.Notification('New Device Added', body_msg.strip(),
                notification_icon)
        n.show()
```

The notification icons are located in

```
/usr/share/icons/Human/scalable/status/
```

and:

```
/usr/share/notify-osd/icons/hicolor/scalable/status/
```

Note that Human is the default icon theme for Ubuntu. This can be seen by running:

```
gconftool-2 -g /desktop/gnome/interface/icon_theme
```

More notification examples can downloaded by getting the latest notify-osd source code:

```
$ bzr get http://bazaar.launchpad.net/~ubuntu-desktop/notify-osd/ubuntu notify-osd
```

The notification system can be turned off on a device by doing:

```
sudo mv /usr/share/dbus-1/services/org.freedesktop.Notifications.service
/usr/share/dbus-1/services/org.freedesktop.Notifications.service.disabled
```

Appropriate use of the notification framework from your application will help create a more uniform desktop experience for your users.

Putting All the Concepts Together

What follows is the complete code for an applet which mounts removable media when it is attached to a device and displays a notification of this to the user. It uses all of the key techniques that were mentioned earlier in this chapter, including D-Bus, notifications and DeviceKit as well as demonstrating the use of gobject, gtk and gconf.

Chapter 4: Application Development

It was written for the karmic release of Netbook Remix and it will not work on Jaunty. This is because the code uses Device Kit, which in karmic replaced HAL as the default hardware abstraction layer.

```python
#!/usr/bin/env python
# -*'- coding: utf-8 -*-

# Copyright (C) 2008 Rodrigo Cesar Lopes Belem
#
# Author: Rodrigo Cesar Lopes Belem <rodrigo.belem@gmail.com>
#
# This program is free software; you can redistribute it and/or modify
# it under the terms of the GNU General Public License as published by
# the Free Software Foundation; either version 2 of the License, or
# (at your option) any later version.
#
# This program is distributed in the hope that it will be useful,
# but WITHOUT ANY WARRANTY; without even the implied warranty of
# MERCHANTABILITY or FITNESS FOR A PARTICULAR PURPOSE.  See the
# GNU General Public License for more details.
#
# You should have received a copy of the GNU General Public License
# along with this program; if not, write to the Free Software
# Foundation, Inc., 59 Temple Place, Suite 330, Boston, MA 02111-1307 USA

import os
import sys
import gtk
import math
import gettext
import pynotify
from gettext import gettext as _

import glob
import gobject

locale_dirs = [
    'locale',
    os.path.join(os.path.sep,'usr','share','locale'),
    os.path.join(os.path.sep,'usr','local','share','locale'),
    ]
DIR = None
for l in locale_dirs:
    if glob.glob(os.path.join(l,'*','*','disk-applet.mo')):
        DIR = l
        break


gettext.bindtextdomain('disk-applet', DIR)
gettext.textdomain('disk-applet')
gettext.install('disk-applet', DIR, unicode=1)

import locale
try:
    locale.setlocale(locale.LC_ALL,'')
```

Chapter 4: Application Development

```python
except locale.Error:
    print 'Unable to properly set locale %s.%s'%(locale.getdefaultlocale())

import dbus
from dbus.mainloop.glib import DBusGMainLoop

DBusGMainLoop(set_as_default=True)


class DiscTray(gtk.StatusIcon):

    def __init__(self):
        gtk.StatusIcon.__init__(self)

        self.set_from_stock(gtk.STOCK_HARDDISK)
        self.set_tooltip('Status icon tooltip')

        self._devices = {}

        # saving notification system availablity
        self._notification_available = pynotify.init('disk-applet')

        try:
            bus = dbus.SystemBus()
        except dbus.DBusException:
            exit(1)

        try:
            bus.add_signal_receiver(self.on_device_added,
                                    'DeviceAdded',
                                    'org.freedesktop.DeviceKit.Disks',
                                    'org.freedesktop.DeviceKit.Disks',
                                    '/org/freedesktop/DeviceKit/Disks')

            bus.add_signal_receiver(self.on_device_removed,
                                    'DeviceRemoved',
                                    'org.freedesktop.DeviceKit.Disks',
                                    'org.freedesktop.DeviceKit.Disks',
                                    '/org/freedesktop/DeviceKit/Disks')

            bus.add_signal_receiver(self._on_device_changed,
                                    'DeviceChanged',
                                    'org.freedesktop.DeviceKit.Disks',
                                    'org.freedesktop.DeviceKit.Disks',
                                    '/org/freedesktop/DeviceKit/Disks')

        except dbus.DBusException:
            exit(1)

        self.connect('activate', self._on_activate_event)
        self.connect('popup-menu', self._on_popup_menu_event)

        # add a signal to detect when a cd is inserted into the drive
        gobject.signal_new('device-is-optical-disc', gtk.StatusIcon,
            gobject.SIGNAL_RUN_LAST, gobject.TYPE_NONE,
            (gobject.TYPE_STRING , gobject.TYPE_BOOLEAN))
```

Chapter 4: Application Development

```python
        # add a signal to detect when a device is mounted, so an action can be
        # performed after that
        gobject.signal_new('device-mounted', gtk.StatusIcon,
            gobject.SIGNAL_RUN_LAST, gobject.TYPE_NONE, (gobject.TYPE_STRING,))
        gobject.signal_new('device-unmounted', gtk.StatusIcon,
            gobject.SIGNAL_RUN_LAST, gobject.TYPE_NONE, (gobject.TYPE_STRING,))

        self.connect('device-is-optical-disc',
                self._on_optical_disc_insert_event)
        self.connect('device-mounted', self._on_device_mounted_event)
        self.connect('device-unmounted', self._on_device_unmounted_event)

        self._populate_device_list()

    def _get_human_readable_size(self, bytes):
        bytes = float(bytes)

        count = int(math.log(bytes) / math.log(1000))
        if count == 1:
            bin_prefix = 'KB'
        elif count == 2:
            bin_prefix = 'MB'
        elif count == 3:
            bin_prefix = 'GB'
        elif count == 4:
            bin_prefix = 'TB'

        return '%01.1f%s' % (bytes/1000**count, bin_prefix)

    def _get_dev_iface(self, object_path, interface):
        bus = dbus.SystemBus()
        proxy_obj = bus.get_object('org.freedesktop.DeviceKit.Disks',
                object_path)
        iface_obj = dbus.Interface(proxy_obj, interface)

        return iface_obj

    def _get_dev_prop(self, device, property):
        props_iface_obj = self._get_dev_iface(device,
                'org.freedesktop.DBus.Properties')

        try:
            res = props_iface_obj.Get('org.freedesktop.DeviceKit.Disks.Device',
                    property)
        except dbus.DBusException:
            res = ''

        return res

    def _get_all_dev_prop(self, device):
        props_iface_obj = self._get_dev_iface(device,
                'org.freedesktop.DBus.Properties')

        return props_iface_obj.GetAll('org.freedesktop.DeviceKit.Disks.Device')

    def _get_dev_methods(self, device):
```

Chapter 4: Application Development

```python
        device_iface_obj = self._get_dev_iface(device,
                'org.freedesktop.DeviceKit.Disks.Device')

        return device_iface_obj

    def _populate_device_list(self):

        bus = dbus.SystemBus()
        disks_obj = bus.get_object('org.freedesktop.DeviceKit.Disks',
                '/org/freedesktop/DeviceKit/Disks')

        devices = disks_obj.EnumerateDevices()

        if devices:
            devices.sort()
            for device in devices:
                self._add_device_to_list(device)

    def _add_device_to_list(self, device):

        if self._get_dev_prop(device, 'device-is-partition') \
                or self._get_dev_prop(device, 'DriveIsMediaEjectable'):
            self._devices[device] = dict(self._get_all_dev_prop(device))

#
# SIGNALS CALLBACKS
#

    def on_device_added(self, device):
        self._add_device_to_list(device)

        device_props = dict(self._get_all_dev_prop(device))

        if not device_props['DeviceIsDrive'] \
                or not device_props['DriveIsMediaEjectable'] \
                and not self._notification_available:
            return

        if device_props['DriveConnectionInterface'] == 'usb':
            notification_icon = 'notification-device-usb'
        elif device_props['DriveConnectionInterface'] == 'firewire':
            notification_icon = 'notification-device-firewire'
        else:
            notification_icon = 'notification-device'

        body_msg = '%s %s %s' % (
                device_props['DriveVendor'],
                device_props['DriveModel'],
                device_props['DriveRevision'])

        n = pynotify.Notification('New Device Added', body_msg.strip(),
                notification_icon)
        n.show()

    def on_device_removed(self, device):
```

Chapter 4: Application Development

```python
        if device not in self._devices:
            return

        device_props = self._devices[device]

        if device_props['DriveConnectionInterface'] == 'usb':
            notification_icon = 'notification-device-usb'
        elif device_props['DriveConnectionInterface'] == 'firewire':
            notification_icon = 'notification-device-firewire'
        else:
            notification_icon = 'notification-device'

        body_msg = '%s %s %s' % (
                device_props['DriveVendor'],
                device_props['DriveModel'],
                device_props['DriveRevision'])

        n = pynotify.Notification('Device Removed', body_msg.strip(),
                notification_icon)
        n.show()

        # removing device from our list
        self._devices.pop(device)

    def _on_device_changed(self, device):
        if device not in self._devices:
            return

        cur_dev_data = self._devices[device]
        new_dev_data = self._get_all_dev_prop(device)

        if new_dev_data['DeviceIsOpticalDisc'] \
                != cur_dev_data['DeviceIsOpticalDisc']:
            self.emit('device-is-optical-disc', device,
                    new_dev_data['DeviceIsOpticalDisc'])
        if new_dev_data['DeviceIsMounted'] != cur_dev_data['DeviceIsMounted']:
            if new_dev_data['DeviceIsMounted']:
                self.emit('device-mounted', device)
            else:
                self.emit('device-unmounted', device)

        # saving the new device data
        self._devices[device] = new_dev_data

    def _on_optical_disc_insert_event(self, status_icon, device, status):
        pass

    def _on_device_mounted_event(self, status_icon, device):
        pass

    def _on_device_unmounted_event(self, status_icon, device):
        pass

    def on_mount_volume(self, menuitem, device):
        if self._devices['DeviceIsMounted']:
```

Chapter 4: Application Development

```python
                return False

        if device['IdLabel']:
            label = device['IdLabel']
        else:
            label = device['IdUuid']

        mount_path = os.path.join('/media', label)

        if self._device['IdType'] == 'vfat':
            mount_options = ['shortname=mixed', 'uid=%s' % os.getuid(),
                    'gid=%s' % os.getgid(), 'shortname=lower', 'dmask=0077',
                    'utf8=1', 'flush']
        else:
            mount_options = []

        try:
            device_iface_obj = self._get_dev_methods(device)
            device_iface_obj.FilesystemMount(self._device['IdType'], fs_args,
                    mount_path)
        except dbus.DBusException:
            return False

        return True

    def on_umount_device(self, menuitem, device):
        try:
            device_iface_obj = self._get_dev_methods(device)
            device_iface_obj.FilesystemUnmount([])
        except dbus.DBusException:
            return False

        return True

    #
    # UI CALLBACKS
    #

    def _on_activate_event(self, status_icon):

        ui_manager = gtk.UIManager()

        ui_manager.add_ui(ui_manager.new_merge_id(), '/', 'popup', 'popup',
                gtk.UI_MANAGER_POPUP, False)

        action_list = []

        if self._devices:

            for device_obj_path in self._devices:

                device = self._devices[device_obj_path]

                if device['IdUsage'] == 'filesystem':

                    if device['IdLabel']:
```

Chapter 4: Application Development

```python
            label = device['IdLabel']
        else:
            label = '%s Filesystem' \
                    % self._get_human_readable_size(
                        device['DeviceSize'])

    uuid = str(device['IdUuid'])
    ui_manager.add_ui(ui_manager.new_merge_id(), '/popup',
            uuid, uuid, gtk.UI_MANAGER_MENU, False)

    action_list.append((uuid, gtk.STOCK_HARDDISK, label,
            None, label))
    # submenu goes here
    if device['DeviceIsMounted']:
        # browse the contents
        ui_manager.add_ui(ui_manager.new_merge_id(),
                '/popup/' + uuid, 'open' + uuid,
                'open' + uuid,
                gtk.UI_MANAGER_MENUITEM, False)
        action_list.append(('open' + uuid, gtk.STOCK_OPEN,
            'Browse', None, 'Browse the device contents',
            self.on_about_event))

        # unmount the device
        ui_manager.add_ui(ui_manager.new_merge_id(),
                '/popup/' + uuid, 'unmount' + uuid,
                'unmount' + uuid,
                gtk.UI_MANAGER_MENUITEM, False)
        action_list.append(('unmount' + uuid, gtk.STOCK_OPEN,
            'UnMount', None, 'UnMount the device',
            self.on_about_event))
    else:
        # unmount the device
        ui_manager.add_ui(ui_manager.new_merge_id(),
                '/popup/' + uuid, 'mount' + uuid,
                'mount' + uuid,
                gtk.UI_MANAGER_MENUITEM, False)
        action_list.append(('mount' + uuid, gtk.STOCK_APPLY,
            'Mount', None, 'Mount the device',
            self.on_about_event))


elif device['DriveIsMediaEjectable']:
    if device['IdLabel']:
        label = device['IdLabel']
    else:
        label = '%s %s %s' % (
                device['DriveVendor'],
                device['DriveModel'],
                device['DriveRevision'])

    ui_manager.add_ui(ui_manager.new_merge_id(), '/popup',
            label, label, gtk.UI_MANAGER_MENU, False)

    action_list.append((label, gtk.STOCK_CDROM, label,
```

Chapter 4: Application Development

```python
                        None, label, self.on_about_event))

                if device['DeviceIsMounted']:
                    # browse the contents
                    ui_manager.add_ui(ui_manager.new_merge_id(),
                            '/popup/' + label, 'open', 'open',
                            gtk.UI_MANAGER_MENUITEM, False)
                    action_list.append(('open', gtk.STOCK_OPEN,
                        'Browse', None, 'Browse the device contents',
                        self.on_about_event))

                    # unmount the device
                    ui_manager.add_ui(ui_manager.new_merge_id(),
                            '/popup/' + label, 'unmount', 'unmount',
                            gtk.UI_MANAGER_MENUITEM, False)
                    action_list.append(('unmount', gtk.STOCK_OPEN,
                        'UnMount', None, 'UnMount the device',
                        self.on_about_event))
                else:
                    # unmount the device
                    ui_manager.add_ui(ui_manager.new_merge_id(),
                            '/popup/' + uuid, 'mount', 'mount',
                            gtk.UI_MANAGER_MENUITEM, False)
                    action_list.append(('mount', gtk.STOCK_APPLY,
                        'Mount', None, 'Mount the device',
                        self.on_about_event))

        else:

            ui_manager.add_ui(ui_manager.new_merge_id(), '/popup',
                    'WithoutDisks', 'WithoutDisks',
                    gtk.UI_MANAGER_MENUITEM, False)
            action_list.append(('WithoutDisks', gtk.STOCK_DELETE,
                _('Without Disks'), None, _('Without Disks'),
                self.on_about_event))


        action_group = gtk.ActionGroup('actions1')
        action_group.add_actions(action_list)

        ui_manager.insert_action_group(action_group, 0)
        ui_manager.ensure_update()

        menu = ui_manager.get_widget('/popup')
        menu.show_all()

        menu.popup(None, None, gtk.status_icon_position_menu, 1,
                gtk.get_current_event_time(), status_icon)

    def _on_popup_menu_event(self, status_icon, button, activate_time):

        action_group = gtk.ActionGroup('actions2')
        action_group.add_actions(
```

Chapter 4: Application Development

```
                [
                    ('Preferences', gtk.STOCK_PREFERENCES, 'Preferences', None,
                        'Preferences', self.on_preferences_event),
                    ('About', gtk.STOCK_ABOUT, 'About', None, 'About',
                        self.on_about_event)
                ])

        ui_string = \
            """<ui>
                    <popup>
                        <menuitem name="Preferences" action="Preferences"/>
                        <separator/>
                        <menuitem name="About" action="About"/>
                    </popup>
                </ui>
            """

        ui_manager = gtk.UIManager()
        ui_manager.add_ui_from_string(ui_string)
        ui_manager.insert_action_group(action_group, 0)

        menu = ui_manager.get_widget('/popup')
        menu.show_all()
        menu.popup(None, None, gtk.status_icon_position_menu, button,
            activate_time, status_icon)

    def on_preferences_event(self, *args):
        pass

    def on_help_event(self, *args):
        pass

    def on_about_event(self, *args):
        pass

    def run(self):
        try:
            gtk.main()
        except KeyboardInterrupt:
            print 'Exiting...'
            exit(0)

if __name__ == '__main__':

    tray_icon = DiscTray()
    tray_icon.run()
```

Summary

A fifth of the world's population will soon have a mobile device and access to the Internet. With that many potential users, an explosion of mobile applications is inevitable and is already happening. This chapter helped you understand the types of technologies that will lead this development.

Application Packaging

There are several reasons why you might want to learn how to package for Ubuntu. First, building and fixing Ubuntu packages is a great way to contribute to the Ubuntu community. It is also a good way to learn how Ubuntu and the applications you have installed work. It is also useful if you need to package an application that is not yet in the archive.

Finally, it is obviously important if you intend to highly customize the default set of applications that come with Ubuntu Mobile.

Background and Important Tools

One of the things that makes Ubuntu such a well respected Linux distribution is its packaging system, which is based on Debian which is seen as one of the most elegant methods of installing, upgrading, and removing software available in the Free Software world.

To become a good packager some familiarity with the following tools is necessary (this list is not exhaustive, but rather a snapshot of some of the applications which we use):

./configure

>Software is generally developed to be used on multiple platforms. Because each of these platforms has different compilers and different include files, there is a need to write Makefiles (see the next entry) and build scripts so that they can work on a variety of platforms. The GNU Project, faced with this problem, devised a set of tools to help with this task. The `configure` script runs a series of tests to determine important information about your machine.

make

>Make is a tool that controls the generation of executables and other non-source files of a program from the program's source files.

Chapter 5: Application Packaging

Make gets its knowledge of how to build your program from a file called a `Makefile`. A makefile is a special file containing shell commands, that you create and name `Makefile`, and which is executed by typing the command `make` while in the same directory as the `Makefile`. When you write a program, you should write a makefile for it, so that it is possible to use `make` to build and install the program.

apt

apt is an acronym for the Advanced Package Tool. This tool has been used extensively throughout the book. The following apt commands are useful for packaging:

apt-cache dump — Shows every package in the cache. This command is useful in combination with the grep pipe such as `apt-cache dump | grep bar` to search for packages whose names or dependencies include bar.

apt-cache policy — Lists the repositories (main/restricted/universe/multiverse) in which a package exists.

apt-cache show — Shows information about a binary package.

apt-cache showsrc — Shows information about a source package.

apt-cache rdepends — Shows reverse dependencies for a package (which packages require the queried one).

apt-rdepends — Recursively lists package dependencies as well as forward build-dependencies.

dpkg

dpkg is a Debian packaging tool that can be used to install, query, and uninstall packages. The following commands are useful for packaging:

dpkg -S — Shows the binary package to which a particular file belongs.

dpkg -l — Shows currently installed packages. This is similar to apt-cache dump but for installed packages.

dpkg -c — Shows the contents of a binary package. It is useful for ensuring that files are installed to the right places.

dpkg -f — Shows the control file for a binary package. It is useful for ensuring that the dependencies are correct.

dpkg -L — Lists files "owned" by package(s).

dpkg -s — Displays package status details.

dpkg-source

A very useful command when working with source packages is:

```
$ dpkg-source -x filename.dsc [output-directory]
```

This extracts a source package. One non-option argument must be supplied, the name of the Debian source control file (`.dsc`). An optional second non-option argument may be supplied to specify the directory to extract the source package to — this must not exist.

Chapter 5: Application Packaging

If no output directory is specified, the source package is extracted into a directory named source-version under the current working directory.

dpkg-scanpackages

dpkg-scanpackages sorts through a tree of Debian binary packages and creates a Packages.gz file, which can be used by a device's software update tool.

dpkg-scansources

dpkg-scanpackages sorts through a tree of Debian source packages and creates a Packages.gz file, which can be used by a device's software update tool.

To use the two preceding commands and to make your custom Debian packages apt-gettable, you need the following files all in the same directory:

- ❑ The binary packages (.deb)
- ❑ The source packages (.orig.tar.gz, .diff.gz, and .dsc)
- ❑ An optional override file

diff

The `diff` program can be used to compare two files and to make patches. A typical example might be `diff -ruN file.old file.new > file.diff`. This command will create a diff (recursively if directories are used) that shows the changes, or "delta" between the two files.

patch

The patch program is used to apply a patch (usually created by diff) to a file or directory. To apply the patch created previously, use `patch -p0 < file.diff`. The -p<num> parameter strips leading slashes from each file found in the patch. For example, if the file name was /src/ui/list.c, setting p0 gives the complete path whereas setting p1 gives src/ui/list.c.

build-essential

This package contains a list of required tools that are considered essential when building software from source.

If you have this build-essential software installed, you only need to install whatever a package specifies as its build-time dependencies to build that package.

devscripts

devscripts contains many scripts that make the package work easier, including debclean, debdiff, and dget.

debclean is used to purge a Debian source tree, debdiff is used to compare two versions of a Debian package to check for added or removed files, and dget downloads Debian source and binary packages.

gnupg

gnupg is a complete and free replacement for PGP used to digitally sign files (including packages).

Chapter 5: Application Packaging

fakeroot

> Gives a fake root environment and enables the running of a command like this (notice the user $ and not the root #):

```
$ dpkg-buildpackage -rfakeroot
```

> It is useful for building packages as a normal user.

lintian

> lintian contains automated checks for many aspects of Debian policy as well as some checks for common packaging errors.
>
> lintian uses an archive directory (called a laboratory), which defaults to /tmp and in this directory it stores information about the packages it examines. It can keep this information between multiple invocations in order to avoid repeating expensive data-collection operations.

pbuilder

> Using pbuilder as a package builder enables you to build the package from within a chroot environment. You can build binary packages without using builder, but you must have all the build dependencies installed on your system first. PBuilder customization is covered in more detail later in this chapter..

Packaging and Using a PPA

With Launchpad's Personal Package Archives (PPA) it is possible to build and then publish binary packages for multiple architectures by uploading source code to the web service. Currently, the limit on the size of a personal archive is 1GB per PPA. If you have multiple PPA's, it is 1GB per PPA. To use the service, your PGP key needs to be uploaded and you also need to have signed the Ubuntu Code of Conduct.

To show packaging in a real-world situation, we will package the ubuntu-golden theme, which will be created in Chapter 7. This will be uploaded to a Launchpad PPA.

> *For more information on Launchpad see Appendix C.*

Create a folder called ubuntu-golden and inside this folder create two other folders called debian and gtk-2.0. Inside the gtk-2.0 folder, place the gtkrc file we created in the theming chapter.

The debian directory is where all the packaging information is stored and it allows us to separate the packaging files from the application source files. Inside this folder, we now need to create the essential files for any Ubuntu source package: changelog, control, copyright, and rules. These are the files needed to create the binary packages (.deb files) from the original source code. We will look at each file in turn.

> *This is just a quick demonstration of some of the files that can make up an Ubuntu package. As such, there is a lot of information about packaging that is missing. For additional information, please read the Packaging Guide at* http://wiki.ubuntu.com/. *Packaging is an art form and if you find after going through this chapter that you have a natural ability or interest in this area, then this guide is the best place to start.*

Chapter 5: Application Packaging

Initial Debianization

If dh_make is not installed, run

```
$ sudo apt-get install dh-make
```

and then run this tool:

```
$ ln -s ubuntu-golden ubuntu-golden-0.1
$ dh_make -createorig -indep
$ ls debian/
changelog dirs init.d.ex menu.ex README.Debian
compat docs init.d.lsb.ex postinst.ex rules
control emacsen-install.ex manpage.1.ex postrm.ex teste.default.ex
copyright emacsen-remove.ex manpage.sgml.ex preinst.ex teste.doc-base.EX
cron.d.ex emacsen-startup.ex manpage.xml.ex prerm.ex watch.ex
```

rules

The rules file is an executable Makefile that has rules for building the binary package from the source packages. A full explanation of a rules file can be found at https://wiki.ubuntu.com/PackagingGuide/Complete#rules.

A default rules file created by `dh_make` looks like:

```
#!/usr/bin/make -f
# -*- makefile -*-
# Sample debian/rules that uses debhelper.
# This file was originally written by Joey Hess and Craig Small.
# As a special exception, when this file is copied by dh-make into a
# dh-make output file, you may use that output file without restriction.
# This special exception was added by Craig Small in version 0.37 of dh-make.
# Uncomment this to turn on verbose mode.
#export DH_VERBOSE=1
configure: configure-stamp
configure-stamp:
        dh_testdir
        # Add here commands to configure the package.

        touch configure-stamp

build: build-stamp

build-stamp: configure-stamp
        dh_testdir

        # Add here commands to compile the package.
        $(MAKE)
        #docbook-to-man debian/teste.sgml > teste.1

        touch $@
```

Chapter 5: Application Packaging

```
clean:
        dh_testdir
        dh_testroot
        rm -f build-stamp configure-stamp

        # Add here commands to clean up after the build process.
        $(MAKE) clean

        dh_clean

install: build
        dh_testdir
        dh_testroot
        dh_prep
        dh_installdirs

        # Add here commands to install the package into debian/teste.
        $(MAKE) DESTDIR=$(CURDIR)/debian/teste install

# Build architecture-independent files here.
binary-indep: install
        dh_testdir
        dh_testroot
        dh_installchangelogs
        dh_installdocs
        dh_installexamples
#       dh_install
#       dh_installmenu
#       dh_installdebconf
#       dh_installlogrotate
#       dh_installemacsen
#       dh_installpam
#       dh_installmime
#       dh_installinit
#       dh_installcron
#       dh_installinfo
#       dh_installwm
#       dh_installudev
#       dh_lintian
#       dh_undocumented
        dh_installman
        dh_link
        dh_compress
        dh_fixperms
#       dh_perl
#       dh_python
        dh_installdeb
        dh_gencontrol
        dh_md5sums
        dh_builddeb

# Build architecture-dependent files here.
binary-arch: install
```

Chapter 5: Application Packaging

```
binary: binary-indep binary-arch
.PHONY: build clean binary-indep binary-arch binary install configure
```

The main thing to notice in our modified rules file is the code in the following block:

```
install -d $(CURDIR)/debian/xfce4-theme-ubuntu-golden/usr/share/themes/
UbuntuGolden/gtk-2.0/
        for file in $(CURDIR)/gtk-2.0/*; do \
        install -c -m 644 $$file $(CURDIR)/debian/xfce4-theme-ubuntu-golden/usr/share/
themes/UbuntuGolden/gtk-2.0/; \
        done
```

It takes our theme and installs it in /usr/share/UbuntuGolden and then changes the permissions to read-only for everyone apart from the owner of the theme. We removed from the default rules file all of the `dh_` commands which were commented out along with dh_installexamples, dh_installman and dh_link, which are not necessary in this case. We also removed the configure-stamp stanza and the Make calls. The full rules file looks like this and notice the various `dh_` commands

```
#!/usr/bin/make -f
# -*- makefile -*-
# Uncomment this to turn on verbose mode.
#export DH_VERBOSE=1
build: build-stamp
build-stamp:
        dh_testdir
        touch build-stamp
clean:
        dh_testdir
        dh_testroot
        rm -f build-stamp
        dh_clean
install: build
        dh_testdir
        dh_testroot
        dh_clean -k
        dh_installdirs
        # install the Gtk+ theme
        install -d $(CURDIR)/debian/xfce4-theme-ubuntu-golden/usr/share/themes/
UbuntuGolden/gtk-2.0/
        for file in $(CURDIR)/gtk-2.0/*; do \
        install -c -m 644 $$file $(CURDIR)/debian/xfce4-theme-ubuntu-golden/usr/share/
themes/UbuntuGolden/gtk-2.0/; \
        done
# Build architecture-independent files here.
binary-indep: build install
        dh_testdir
        dh_testroot
        dh_installchangelogs
        dh_installdocs
        dh_strip
        dh_compress
        dh_fixperms
        dh_installdeb
```

```
        dh_gencontrol
        dh_md5sums
        dh_builddeb

# Build architecture-dependent files here.
binary-arch: install
binary: binary-indep binary-arch
.PHONY: build clean binary-indep binary-arch binary install configure
```

The `dh_` *commands mentioned earlier are part of the* `debhelper` *suite of tools. These help with repetitive tasks when writing a* `rules` *file. For example, the command* `dh_testdir` *tries to make sure that you are in the correct directory when building a debian package. For more information on all of the tools available in the suite look at the debhelper manpages.*

There are other ways to make rules files, such as by using debhelper 7 and cdbs. For more information on these tools, look at `http://manpages.ubuntu.com/manpages/karmic/man7/debhelper.7.html` and `http://build-common.alioth.debian.org/cdbs-doc.html`.

changelog

The changelog file is a listing of the changes made in each version. It has a specific format that gives the package name, version, distribution, changes, and who made the changes at a given time. If you have a GPG key, make sure to use the same name and e-mail address in changelog as you have on the key. Create a changelog of

```
xfce4-theme-ubuntu-golden (0.1jaunty1) jaunty; urgency=low
  * Initial Release.
 -- Ian Lawrence <debs@ianlawrence.info>  Sat, 28 Feb 2009 19:45:07 +0000
```

A changelog can be edited using a tool called `dch`. In the root source directory, type `$ dch -e` to edit the changelog in your preferred editor. To increment the changelog for example when making a new release run `$ dch -i`. To append a new comment to the current version entry, run `$ dch -a`.

> **To set a default editor run:**
>
> $ update-alternatives --config-editor

control

Control data is stored in a control file and it is used for both source and binary packages.

The control file shown below is the control file from our source package. We can tell that this is for a source package because it is separated into two paragraphs, the first of which always refers to the source package itself and the second and subsequent paragraphs to the binary package(s) created from our source.

Chapter 5: Application Packaging

```
Source: xfce4-theme-ubuntu-golden
Section: graphics
Priority: optional
Maintainer: Ian Lawrence < debs@ianlawrence.info >
Build-Depends: debhelper ( > = 7.0.0)
Standards-Version: 3.8.0
Homepage: http://ianlawrence.info

Package: xfce4-theme-ubuntu-golden
Architecture: all
Depends: gtk2-engines-murrine
Description: A golden theme for Gtk+ 2.0
This package contains the Ubuntu Golden package for gtk2.
```

Packages can state that they have relationships to other packages in the control file. In the example above the Depends stanza shows a binary dependency on gtk2-engines-murrine, which means that the xfce4-theme-ubuntu-golden package will not be configured unless gtk2-engines-murrine has already been correctly configured on the system.

After the changelog has been written, it is time to think about which copyright will cover the package. This will most often be decided upstream but if you have written the software then this choice is yours.

copyright

Your copyright file must contain the following information:

- ❏ The author(s) name
- ❏ The year(s) of the copyright
- ❏ The package license(s)(optional)
- ❏ The URL to the upstream source

Every file that contains a different license from the main license must be mentioned along with the license itself. If licensing under the GPL, in the source package there needs to be a section with a header which looks like:

```
---
This program is free software; you can redistribute it and/or modify
it under the terms of the GNU General Public License as published by
the Free Software Foundation; either version 2 of the License, or
(at your option) any later version.

This program is distributed in the hope that it will be useful,
but WITHOUT ANY WARRANTY; without even the implied warranty of
MERCHANTABILITY or FITNESS FOR A PARTICULAR PURPOSE.  See the
GNU General Public License for more details.
---
```

Make sure that there is a reference to distribute the program under a certain license (whether GPL or not)–simply including a reference to the GPL (as shown below) means that in reality no copyright license has been granted. Be careful.

Chapter 5: Application Packaging

```
                ---
                    GNU GENERAL PUBLIC LICENSE
                       Version 2, June 1991

 Copyright (C) 1989, 1991 Free Software Foundation, Inc.
     59 Temple Place, Suite 330, Boston, MA  02111-1307  USA
                ---
```

A template (which was discussed on the debian-legal mailing list) for a copyright file looks like

```
 Authors: Ian Robert Lawrence, Rodrigo Cesar Lopes Belem
 Copyright 2007,2008 Rodrigo Cesar Lopes Belem
              2009  Ian Robert Lawrence

 This program is free software; you can redistribute it and/or modify
 it under the terms of the GNU General Public License as published by
 the Free Software Foundation; either version 2 of the License, or
 (at your option) any later version.

 This program is distributed in the hope that it will be useful,
 but WITHOUT ANY WARRANTY; without even the implied warranty of
 MERCHANTABILITY or FITNESS FOR A PARTICULAR PURPOSE.  See the
 GNU General Public License for more details.

 You should have received a copy of the GNU General Public License with
 the Debian GNU/Linux distribution in file /usr/share/common-licenses/GPL;
 if not, write to the Free Software Foundation, Inc., 59 Temple Place,
 Suite 330, Boston, MA  02111-1307  USA

 On Debian systems, the complete text of the GNU General Public
 License, version 2, can be found in /usr/share/common-licenses/GPL-2.
```

Other Debian Files

compat — Just a version number for the debhelper scripts

install — This file is used to install files in the package

dirs — This file lists which directories will be created in the package

docs — Lists which documentation files or directories will be installed

watch — Monitors upstream for new version releases of the source package

init — init script

pam — pam files that can be installed in /etc/pam.d

links — create links to files inside the package

examples — install examples to the package

Chapter 5: Application Packaging

Building the Package

Now, with the files in place it is possible to actually build the package. Move into the `ubuntu-golden` folder and issue the command

```
$ debuild -S -sa
```

This will build and sign the deb. We run the `debuild` command because our theme is a brand new package with no existing version in Ubuntu's repositories and so dput (covered later) will upload the .orig.tar.gz file to the PPA.

For more information about `debuild` *and PPAs, look at* `https://help.launchpad.net/Packaging/PPA`.

Uploading to a PPA

To upload the new xfce4-theme-ubuntu-golden Debian package to a PPA begin by installing dput:

```
$ sudo apt-get install dput
```

dput is the tool you use to upload your source package to Launchpad. It uploads the following files:

- .dsc
- .changes
- .diff.gz
- And optionally, .orig.tar.gz (if you used `debuild -S -sa` to build the package)

To upload to a PPA, use:

```
$ dput ppa:ianlawrence/ppa <changesfile>
```

This, in the case of our package, will be:</Para>

```
$ ppa:ianlawrence/ppa xfce4-theme-ubuntu-golden_0.1jaunty1_source.changes
```

This outputs:

```
Checking Signature on .changes
gpg: Signature made Sat 28 Feb 2009 20:41:45 GMT using DSA key ID 1C1EABFF
gpg: Good signature from "Ian Lawrence <debs@ianlawrence.info>"
gpg:                aka "Ian Lawrence <root@ianlawrence.info>"
Checking Signature on .dsc
gpg: Signature made Sat 28 Feb 2009 20:41:37 GMT using DSA key ID 1C1EABFF
gpg: Good signature from "Ian Lawrence <debs@ianlawrence.info>"
gpg:                aka "Ian Lawrence <root@ianlawrence.info>"
Good signature on xfce4-theme-ubuntu-golden_0.1jaunty1.dsc.
Uploading to my-ppa (via ftp to ppa.launchpad.net):
  gxfce4-theme-ubuntu-golden_0.1jaunty1.dsc: done.
Successfully uploaded packages.
Not running dinstall.
```

Chapter 5: Application Packaging

When the package has been built by the autobuilders, an e-mail is sent to inform the user of this fact. This is an e-mail from an upload of a GPS daemon called gypsy, which was packaged as an example for REVU (see below) for possible inclusion into the Ubuntu Mobile distribution

```
Accepted:
 OK: gypsy_0.5.orig.tar.gz
 OK: gypsy_0.5-2jaunty1.diff.gz
 OK: gypsy_0.5-2jaunty1.dsc
    -> Component: universe Section: devel

Format: 1.8
Date: Sat, 28 Feb 2009 19:45:07 +0000
Source: gypsy
Binary: gypsy gypsy-dbg libgypsy0 libgypsy0-dbg libgypsy-dev libgypsy-doc
Architecture: source
Version: 0.5-2jaunty1
Distribution: jaunty
Urgency: low
Maintainer: Ross Burton <ross@debian.org>
Changed-By: Ian Lawrence <debs@ianlawrence.info>
Description:
 gypsy—GPS multiplexing daemon
 gypsy-dbg—GPS multiplexing daemon (debug files)
 libgypsy-dev—GPS daemon client library (development files)
 libgypsy-doc—GPS multiplexing daemon (documentation)
 libgypsy0—GPS multiplexing daemon client library
 libgypsy0-dbg—GPS multiplexing daemon client library (debug files)
Changes:
 gypsy (0.5-2jaunty1) jaunty; urgency=low
 .
  * PPA Rebuild
```

Users will now be able to add the PPA to their sources list by doing

```
deb http://ppa.launchpad.net/ianlawrence/ppa/ubuntu jaunty main
deb-src http://ppa.launchpad.net/ianlawrence/ppa/ubuntu jaunty main
```

and they can then install the packages which are available in the PPA using apt.

If you added another PPA to your launchpad account or team account, you must add the following to ~/.dput.cf :

```
[<ppa alias name>]
fqdn = upload.launchpad.net
method = ftp
incoming = ~<launchpad username>/ubuntu/<other ppa name>
login = anonymous
$ dput <ppa alias name> <changes file>
```

REVU

As mentioned previously, it is a good idea to submit any packages for review. REVU (http://revu.ubuntuwire.com) is a web-based tool that gives people who have worked on packages a chance to show them to Ubuntu Developers so that these more experienced developers can review them for potential inclusion into Ubuntu.

Upload to REVU as follows:

```
$ dput revu gypsy_0.5-2jaunty1_source.changes
```

which gives the acceptance e-mail with information about the package:

```
A new package has been accepted into REVU: gypsy

Package uploaded at: 2009-02-28, 22:42
Package was uploaded by: ianlawrence
REVU URL: http://revu.ubuntuwire.com/details.py?upid=5306
```

Subsequent e-mails may arrive from experienced developers — often with some hints about how to improve the package itself. For example

```
Comment for package gypsy (not advocating)
Number of Advocates: 0
Package uploaded at: 2009-02-28, 22:42
Package uploaded by: ianlawrence
REVU URL: http://revu.ubuntuwire.com/details.py?upid=5306
persia wrote:
1) Sections are a mess
2) Why restrict the architecture
3) /etc/dbus-1/event.d/ is deprecated: use LSB init headers
4) changelog is traditionally truncated for initial upload (and version is odd)
5) Upstream has a newer version available
```

This is a good way to get feedback on package work and a great way to gain experience with packaging.

RFA Packages

There may be packages which the current maintainer has put up for adoption (Request For Adoption). With the devscripts package (mentioned earlier in the chapter) installed, run:

```
$ wnpp-alert
```

This command will print out a list of all packages on the device that are orphaned. The list of orphaned packages will look something like the following:

```
RFH 479951 kvm — Full virtualization on x86 hardware (Need help with ia64, ppc and
s390)
```

If a package is listed that is important to you, send an e-mail to the current maintainer asking if they are willing to let you maintain it. To find the current maintainer, run the following:

```
$ apt-cache show kvm
```

which shows the Original-Maintainer field.

Chapter 5: Application Packaging

Creating Your Own Repository

In some cases, using a PPA is not satisfactory and it is more convenient to have the packages available locally. This is particularly true when building a remix or customization of Ubuntu. If this is the case, there are two ways to create a local repository. The first is the "simple" approach; the second uses a tool called reprepro, which helps automate the process.

Simple Repository

Simple repositories do not organize packages in subdirectories. They only create a list of packages, called Packages.gz, by scanning the directory.

The commands that are used to build this type of repository are `dpkg-scanpackages` and `dpkg-scansource`. The following is an example of the use of these commands:

```
dpkg-scanpackages binary /dev/null | gzip -9c > Packages.gz
dpkg-scansorces binary /dev/null | gzip -9c > Sources.gz
```

Automatic Repository

The automatic repository is the way that large repositories, such as Debian and Ubuntu, are built. This repository format allows access from clients from many platforms in a way in which files will not be duplicated. Such a structure is called a pool.

There are several tools that can be used to build this type of repository. One tool is called reprepro. It manages .deb packages and their related files such as .dsc, .diff, .tar.gz, and .udeb.

Setting Up a Repository

First of all, you will need to create the initial directory structure. It should look like this:

```
$ mkdir /path/to/your/repository/ubuntu
$ mkdir /path/to/your/repository/ubuntu/conf
$ mkdir /path/to/your/repository/ubuntu/incoming
```

This creates the reprepro config file. Your distributions file should look like this:

```
Codename: karmic
Suite: karmic
Components: main
Architectures: i386 source
Description: Repo for book
```

This file should go to `/path/to/your/repository/ubuntu/conf`.

Chapter 5: Application Packaging

Adding Packages to a Repository

To add packages to a repository, use the following:

```
$ reprepro -Vb . include jaunty name_of_file.changes
```

Packages can also be added by doing:

```
reprepro -b  includedeb karmic some-package.deb
```

When working on an older distribution (for example, hardy) you may find that some application functionality from a more recent distribution (for example, intrepid) would be useful. In this case it is necessary to "backport" the application to the older distribution.

Removing Packages From a Repository

Packages can be removed with:

```
$ reprepro -Vb /path/to/repository/ubuntu remove karmic some-package.deb
```

Backporting KVM

Add the sources for intrepid to sources.list, like this:

```
deb-src http://archive.ubuntu.com/ubuntu/ intrepid main universe restricted multiverse
```

Comment out temporarily the other deb-src lines in the sources.list. Next, run the following:

```
$ apt-get source kvm
```

Move into the unpacked folder (in this case, kvm-72+dfsg) and run the following:

```
$ debuild
```

If this gives build dependencies such as the following:

```
dpkg-checkbuilddeps: Unmet build dependencies: quilt (>= 0.40) uuid-dev libsdl1.2-dev libasound2-dev libgnutls-dev nasm texi2html bcc iasl
dpkg-buildpackage: warning: Build dependencies/conflicts unsatisfied; aborting.
```

you need to add this:

```
$ sudo apt-get build-dep kvm
```

Then run `debuild` again. This will create the .deb in the parent folder.

PBuilder

While it is possible to build packages the way we have shown earlier, you must already have all the build dependencies of the package that you are building installed on your running system.

A tool called `pbuilder` can help with this. `pbuilder` checks the build dependencies automatically and it does this because the package is built within a minimal Ubuntu installation with the build dependencies downloaded according to the Debian/control file.

The primary aim of `pbuilder` is different from other auto-building systems in Debian in that it does not try to build as many packages as possible. It does not try to guess what a package needs, and in most cases it tries the worst choice of all if there is a choice to be made.

Configuring PBuilder

It can be a good idea to configure `pbuilder`. There are many ways to do this. In the /etc directory is the main configuration file for `pbuilder` called `pbuilderrc`.

Edit this file to look as follows.

```
BASETGZ=/var/cache/pbuilder/base.tgz
#EXTRAPACKAGES=gcc3.0-athlon-builder
#export DEBIAN_BUILDARCH=athlon
BUILDPLACE=/var/cache/pbuilder/build/
MIRRORSITE=http://archive.ubuntu.com/ubuntu
USEPROC=yes
USEDEVPTS=yes
USEDEVFS=no
BUILDRESULT=/var/cache/pbuilder/result/
# specifying the distribution forces the distribution on "pbuilder update"
DISTRIBUTION=jaunty
# specifying the components of the distribution (default is "main")
#COMPONENTS="main restricted universe multiverse"
#specify the cache for APT
APTCACHE="/var/cache/pbuilder/aptcache/"
APTCACHEHARDLINK="yes"
REMOVEPACKAGES=""
#HOOKDIR="/usr/lib/pbuilder/hooks"
HOOKDIR=""
# make debconf not interact with user
export DEBIAN_FRONTEND="noninteractive"
DEBEMAIL=""
# for pbuilder debuild (sudo -E keeps the environment as-is)
BUILDSOURCEROOTCMD="fakeroot"
PBUILDERROOTCMD="sudo -E"
# command to satisfy build-dependencies; the default is an internal shell
# implementation which is relatively slow; there are two alternate
# implementations, the "experimental" implementation,
# "pbuilder-satisfydepends-experimental", which might be useful to pull
# packages from experimental or from repositories with a low APT Pin Priority,
# and the "aptitude" implementation, which will resolve build-dependencies and
# build-conflicts with aptitude which helps dealing with complex cases but does
```

Chapter 5: Application Packaging

```
# not support unsigned APT repositories
PBUILDERSATISFYDEPENDSCMD="/usr/lib/pbuilder/pbuilder-satisfydepends"
#Command-line option passed on to dpkg-buildpackage.
#DEBBUILDOPTS="-IXXX -iXXX"
DEBBUILDOPTS=""
#APT configuration files directory
APTCONFDIR=""
# the username and ID used by pbuilder, inside chroot. Needs fakeroot, really
BUILDUSERID=1234
BUILDUSERNAME=pbuilder
# BINDMOUNTS is a space separated list of things to mount
# inside the chroot.
BINDMOUNTS=""
# Set the debootstrap variant to 'buildd' type.
DEBOOTSTRAPOPTS[0]=' — variant=buildd'
# or unset it to make it not a buildd type.

# unset DEBOOTSTRAPOPTS
# Set the PATH I am going to use inside pbuilder: default is "/usr/sbin:/usr/bin:/sbin:/bin:/usr/X11R6/bin"
export PATH="/usr/sbin:/usr/bin:/sbin:/bin:/usr/X11R6/bin"
# SHELL variable is used inside pbuilder by commands like 'su'; and they need sane values
export SHELL=/bin/bash
# The name of debootstrap command.
DEBOOTSTRAP="debootstrap"
# default file extension for pkgname-logfile
PKGNAME_LOGFILE_EXTENTION="_$(dpkg — print-architecture).build"
# default PKGNAME_LOGFILE
PKGNAME_LOGFILE=""
```

This provides some sane configuration defaults for `pbuilder`.

It is now possible to script pbuilder using a bash script. The following script allows a Debian package to be created for *any* Ubuntu distribution (the name is passed as a command line parameter to pbuilder). Change the /home/<user>/.pbuilderrc file to look like the code below. This code is for illustration only. Please download this file from our code bundle on the Wrox site.

```
#USE_SYSTEM_DIST=1
#DEFAULT_DIST="hardy"
#OTHERMIRROR="deb http://localhost:8000/test/ ./|deb http://localhost:8000/extra/ hardy main|"
UBUNTU_UPDATES=1
UBUNTU_SECURITY=1
UBUNTU_BACKPORTS=1
UBUNTU_PROPOSED=0
DEBIAN_SECURITY=1
UBUNTULIST="jaunty intrepid hardy karmic"
DEBIANLIST="stable testing unstable sid experimental etch lenny"
UBUNTU_MIRROR="http://br.archive.ubuntu.com/ubuntu/"
DEBIAN_MIRROR="http://ftp.br.debian.org/debian/"
if [ -z "${DIST}" ]; then
```

Chapter 5: Application Packaging

```
            if [ -n "${DEFAULT_DIST}" ]; then
                DIST="${DEFAULT_DIST}"
            elif [ "${USE_SYSTEM_DIST}" ]; then

                : ${DIST:="$(lsb_release - short - codename)"}
            else

                if [ -r "debian/changelog" ]; then
                    DIST=$(dpkg-parsechangelog | awk '/^Distribution: / {print $2}')
                    # Use the unstable suite for debian experimental packages.
                    if [ "${DIST}" == "experimental" ]; then
                        DIST="unstable"
                    fi
                fi
            fi
fi

if [ -n "$(echo ${UBUNTULIST[@]} | grep ${DIST})" ]; then

    COMPONENTS="main restricted universe multiverse"

    MIRRORSITE="${UBUNTU_MIRROR}"

    # Defining which repositories we will use in ubuntu

    if [ "${UBUNTU_UPDATES}" ]; then

        OTHERMIRROR="${OTHERMIRROR}deb ${UBUNTU_MIRROR} ${DIST}-updates
${COMPONENTS}|"

    fi

    if [ "${UBUNTU_SECURITY}" ]; then

        OTHERMIRROR="${OTHERMIRROR}deb ${UBUNTU_MIRROR} ${DIST}-security
${COMPONENTS}|"

    fi

    if [ "${UBUNTU_BACKPORTS}" ]; then

        OTHERMIRROR="${OTHERMIRROR}deb ${UBUNTU_MIRROR} ${DIST}-backports
${COMPONENTS}|"

    fi
    if [ "${UBUNTU_PROPOSED}" ]; then

        OTHERMIRROR="${OTHERMIRROR}deb ${UBUNTU_MIRROR} ${DIST}-proposed
${COMPONENTS}|"

    fi
```

Chapter 5: Application Packaging

```
        elif [ -n "$(echo ${DEBIANLIST[@]} | grep ${DIST})" ]; then

            echo "Using a debian pbuilder environment because DIST is ${DIST}"

            COMPONENTS="main contrib non-free"

            MIRRORSITE="${DEBIAN_MIRROR}"

            # Defining which repositories we will use in debian

            if [ "${DEBIAN_SECURITY}" ]; then

                OTHERMIRROR="${OTHERMIRROR}deb ${DEBIAN_MIRROR} ${DIST}-security
${COMPONENTS}|"

            fi

        fi

    BASETGZ="`dirname ${BASETGZ}`/${DIST}-base.tgz"

    #BASETGZ="/var/cache/pbuilder/${DIST}-base.tgz"

    DISTRIBUTION="${DIST}"

    BUILDRESULT="/var/cache/pbuilder/${DIST}/result/"

    APTCACHE="/var/cache/pbuilder/${DIST}/aptcache/"
```

By configuring your `pbuilder` this way, you have the capability to build packages for different release versions, distributions, and architectures.

To do this, first create the pbuilder:

```
$ sudo DIST=karmic pbuilder create
```

and then pass the release you wish to build for along with the .dsc of the package to the pbuilder like this

```
$ sudo DIST=karmic pbuilder build package_version.dsc
```

So to build the first Hardy version of the Ubuntu Mobile guide (which is available at https://edge.launchpad.net/~ianlawrence/+archive/ppa) for the Jaunty release, run the following:

```
$ sudo DIST=jaunty pbuilder build mobileguide_0.1-2.dsc
```

Performing Actions on PBuilder

Along with the `create` command shown previously, there are several other simple commands for operation such as `pbuilder update` and `pbuilder build`.

Chapter 5: Application Packaging

These commands are covered in a little more detail next.

Creating a Distribution Environment

The command `pbuilder create` will create a base chroot image tar-ball (`base.tgz`). All other commands will operate on the resulting `base.tgz`. The distribution code-name needs to be specified with the `distribution` command-line option.

`debootstrap` is used to create the bare minimum Debian installation, and then `build-essential` packages are installed on top of the minimum installation using `apt-get` inside the chroot.

Use `pbuilder` to `create` a package like this:

```
$ sudo DIST=karmic pbuilder create
```

Building a Package to a Specific Release

To `build` a package inside the chroot, use the following:

```
$ sudo DIST=karmic pbuilder build whatever.dsc
```

`pbuilder` will extract base.tgz to a temporary working directory, enter the directory with chroot, satisfy the build-dependencies inside chroot, and build the package. The built packages will be moved to a directory specified with the `buildresult` command-line option.

Updating the PBuilder Environment

Add the following code:

```
$ pbuilder update
```

It will `update` the base.tgz file. It will extract the chroot, invoke `apt-get update` and `apt-get dist-upgrade` inside the chroot, and recreate base.tgz (the base tar-ball).

Using pdebuild

Common packaging workflow is the one shown in the "Backporting KVM" section. A developer may try to do `debuild` and build a package inside a Debian source directory:

- ❏ `pdebuild` allows a similar workflow with packages built inside the chroot with checks that the current source tree will build happily.
- ❏ `pdebuild` calls `dpkg-source` to build the source packages, and then invokes `pbuilder` on the resulting source package. However, unlike `debuild`, the resulting deb files will be found in the `buildresult` directory.

A slightly different mode of operation is available in `pdebuild` since version 0.97. `pdebuild` usually runs `Debian/rules clean` outside of the chroot; however, it is possible to change the behavior to run it inside the chroot with — `use-pdebuild-internal`. It will try to bind mount the working directory inside chroot, and run `dpkg-buildpackage` inside this.

You use pdebuild like this:

Chapter 5: Application Packaging

```
sudo pdebuild — use-pdebuild-internal
```

Configuring Actions

Hook scripts are used when you want to perform some action at a determined moment during `pbuilder` execution. These actions can be scripted, too.

A hook directory can be defined by the variable HOOKDIR. This can be set in the command line as follows:

```
$ sudo HOOKDIR=/path/to/hookdir pbuilder -build file.dsc
```

You can also set it as follows:

```
$ sudo pbuilder -hookdir /path/to/hookdir -build file.dsc
```

The hook script names must be in the form X<digit><digit><whatever-else-you-want> much like boot scripts. Here is the example `c10shell` hook script, which will invoke a shell if the build fails.

```
#!/bin/bash
# invoke shell if build fails.
apt-get install -y — force-yes vim less
cd /tmp/buildd/*/debian/.
/bin/bash < /dev/tty > /dev/tty 2> /dev/tty
```

Additional Hook Manipulation with PBuilder

Some of the most useful hook manipulation commands we have found are as follows. They start with letters, the first of which is A.

```
A<digit><digit><whatever-else-you-want>
```

This is executed before a build starts — after unpacking the build system, and unpacking the source, and satisfying the build-dependency.

B is used like this:

```
B<digit><digit><whatever-else-you-want>
```

This is executed after the build system finishes building successfully and before copying back the build result.

C is used like this:

```
C<digit><digit><whatever-else-you-want>
```

This is executed after build failure and before cleanup.

D is used like this:

```
D<digit><digit><whatever-else-you-want>
```

Chapter 5: Application Packaging

This is executed before unpacking the source inside the chroot and after setting up the chroot environment.

Hook Script Resource

With Bazaar installed (`sudo apt-get install bzr`), check out the following:

```
bzr branch lp:~kubuntu-members/pbuilder/pbuilder-hooks
```

This contains some useful example scripts. These include the useful `D10aptupdate`, which runs apt-get update before proceeding further, and `B10list-missing`, which lists missing files.

Mount Bind a Package Repository for Use with PBuilder

Bind mounting directories is useful for many tasks. One of the most useful things is to mount bind a directory which is a package repository:

```
BINDMOUNTS="/var/cache/pbuilder/result"
# cat /var/cache/pbuilder/hooks/D70results
#!/bin/sh
cd /var/cache/pbuilder/result/
/usr/bin/dpkg-scanpackages . /dev/null > /var/cache/pbuilder/result/Packages
/usr/bin/apt-get update
```

This way, you can use deb file:/var/cache/pbuilder/result in the `OTHERREPOSITORIES` variable.

Ubuntu Policy

The Ubuntu policy document available at `http://people.canonical.com/~cjwatson//ubuntu-policy/policy.html/` describes "the policy requirements for the Ubuntu distribution." It also discusses policy issues that relate to packages, such as the technical requirements that each package must satisfy to be included in the Ubuntu distribution. It is modeled closely on the Debian policy document.

This document is extremely important. Please read it. One important thing to know is how the archive of packages in Ubuntu is divided up into categories.

Categories

The categories are main, restricted, universe, and multiverse, and each has different requirements in order for a package to be included in them.

- **Main** — Packages in main must not require a package outside of *main* for compilation or execution (thus, the package must not declare a "Depends," "Recommends," or "Build-Depends" relationship on a non-*main* package).

- **Restricted** — Source code may not be available for packages in restricted and modifications may also not be permitted.

- **Universe** — Packages in this category are not supported either by Ubuntu Developers or Canonical. Furthermore, packages must not require a package outside of *main* and *universe* for compilation or execution (thus, the package must not declare a "Depends," "Recommends," or "Build-Depends" relationship on a non-*main* and non-*universe* package).

❑ **Multiverse** — Packages must be placed in *multiverse* if source code is not available, modifications are not permitted, or if they are encumbered by patents or other legal issues that make their distribution problematic.

Every package must also provide a copyright file and distribution license in the file /usr/share/doc/package/copyright.

Sections

The preceding categories are further broken down into sections. These sections are specified in the control file of the Debian package. They are useful for organizing packages into logical blocks. The sections that are available include admin, graphics, web, X11, and metapackages. The full list is available in the policy document itself. The use of metapackages is covered in Chapter 11, "Putting It All Together."

The maintainers of the Ubuntu archive may override the section value to ensure the consistency of the Ubuntu distribution. Consequently, it is important to include this information in the package.

Summary

This chapter shows some basic tools used when packaging and covers how to create a simple archive with the files necessary to create a Debian package. It then shows you how to set up a local repository for packages. You also see how to use and customize pbuilder along with other custom scripts, which can be run before, during, or after a pbuilder run. Finally, we discussed the Ubuntu policy document and how software is categorized in the repositories.

Application Selection

According to some observers, the Mobile Internet Device category is rapidly emerging. A recent study by ABI Research shows that this category will appeal both as tools and toys to a "wide variety of people." The report also forecasts that these products will see a dramatic growth in popularity over the next five years, with worldwide shipments rising from under 3.5 million in 2008 to nearly 90 million in 2012.

According to Strategy Analytics, another research firm, sales of Mobile Internet Devices are expected to exceed $17 billion worldwide annually by 2014.

> *More about the business potential of mobile devices can be found in Chapter 12.*

It may benefit OEMs to try to identify "categories of users" within this "wide variety of people" category in order to tailor the applications that they offer on devices for each targeted market segment. This will mean selecting applications and device settings that make the best use of a mobile device for particular groups of users.

Business Users

This is a large and profitable market segment that many OEMs target. Although it may be true that the line between business users and consumers is becoming less differentiated, application providers seem to be focusing on how to integrate their software with MIDs and netbooks to deliver a consistent experience to enterprise users.

Many firms are facing budgetary pressures and need to both become more efficient and cut spending. Solutions such as hosting office applications and documents in the cloud with the device running a highly customized version of Ubuntu Mobile are increasingly attractive to companies. Even so, having a mobile device used in this way does raise more security concerns than if the employee were, for example, tethered to a workstation at the office.

Chapter 6: Application Selection

A factor therefore driving the adoption of mobile devices in corporate settings is security. Password protection, remote wipe capability, and physical device tracking are the top three factors cited by many companies that are contemplating equipping employees with mobile devices.

Ubuntu comes with a feature to create an encrypted private directory for the default user. You can set this up on a device as follows:

```
$ sudo apt-get install ecryptfs-utils
```

Then run the following:

```
$ ecryptfs-setup-private
```

sudo, *which allows temporary superuser privileges, should not be used for this operation.*

Enter a login passphrase, which is the password used to log in to the device.

Choose a mount passphrase.

Enter the mount passphrase (again).

The ecryptfs-setup-private script creates the /home/*user*/Private directory and tests that the encrypted mount works as expected.

After rebooting the device, there will be an encrypted directory in /home/*user*/Private where all confidential company information can be stored.

> *Automatic, password-less logins such as the one on Ubuntu MID mean that the ~/Private directory is not mounted by default. This is by design and means that a password needs to be entered when clicking on the file "Access Your Private Data" in the ~/Private directory. After this is done, the encrypted data is available to the user.*
>
> *A different passphrase should be used to encrypt the mount; this is only needed if it is necessary to manually recover data.*

Documents

Business needs word processing software to create letters and documents, spreadsheet software for business analysis, and calculation and presentation software for important meetings.

OpenOffice, is a free cross-platform office application suite and is compatible with Microsoft Office, so it is easy to exchange documents with friends and colleagues who use Windows and Microsoft Office.

Chapter 6: Application Selection

A Practical Example

OpenOffice can, however, be fairly resource-intensive, which can be a problem on a mobile device. Because of this, the Ubuntu MID team wrote an application called Trebuchet (treb) which is a file handler for displaying office files.

treb is an interface that reads documents created by Microsoft Office applications. It does this by using OpenOffice as a filter to convert the file to PDF and launching Evince to see the result. After Evince is closed, the temporary PDF file is deleted and treb itself is closed.

treb comes as part of the MID install, but it can also be installed using apt-get; it can be run from the command line to present the GUI shown in Figure 6-1.

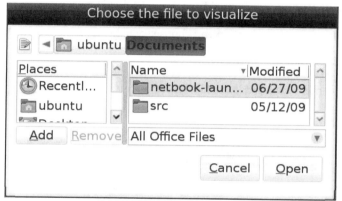

Figure 6-1

```
$ apt-get install treb
$ treb
```

Multimedia Users

Vendors offering devices for the Multimedia Users group have a preexistent and ready-made market growing out of the current market for smart phones. Although multimedia is most often associated with images, text, and music, it can also mean application programs that integrate data from a broad spectrum of independent sources. Probably the best known and most popular application that does this for mobile devices is Canola. It is built using Python and the Enlightenment Foundation Libraries (EFL), and it uses a plug-in system to enable YouTube, Flicker, and more. There are Google Summer of Code 2009 projects approved to create plug-ins for Twitter, Remember the Milk, and even an IM client.

> *EFL is covered in more detail in Chapter 4.*

Another available solution is a media center called Entertainer. This aims to be a simple and easy-to-use application for the Gnome and XFCE desktop environments. Entertainer is written completely in Python and it uses GStreamer's multimedia framework for multimedia playback. The user interface is implemented with the Clutter library.

> *Clutter is covered in more detail in Chapter 4.*

131

Chapter 6: Application Selection

A Practical Example

Entertainer also includes a private video recorder, which means that it can be used to watch Live TV, and it records shows for later viewing. To install entertainer, add the entertainer PPA to /etc/apt/sources.list

```
deb http://ppa.launchpad.net/entertainer-releases/ppa/ubuntu jaunty main
deb-src http://ppa.launchpad.net/entertainer-releases/ppa/ubuntu jaunty main
```

and then do:

```
$ sudo apt-get update && sudo apt-get install entertainer
```

Also add the GPG key to your software sources authentication (System ⇨ Administration ⇨ Software Sources ⇨ Authentication). Select "Import Key File..." and add the key found on the entertainer page.

Then run the content manager application and choose some media folders on the device or feed URLs:

```
$ entertainer-content-manager
```

The interface to the content manager looks like Figure 6-2.

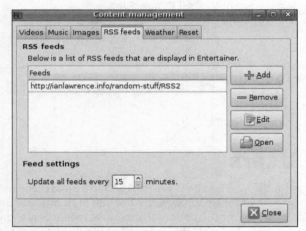

Figure 6-2

Chapter 6: Application Selection

Next, run the front end:

```
$ entertainer
```

Entertainer on the Ubuntu Netbook Remix looks like Figure 6-3.

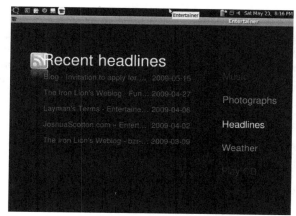

Figure 6-3

To exit Entertainer, press the Q key. Other useful key commands are explained next.

Useful Keybindings in the Entertainer GUI

Entertainer is controlled with the keyboard, as follows:

- **F** — Toggle full-screen on/off.
- **P** — Toggle pause/play when video or audio is playing.
- **S** — Stop playback.
- **H** — Navigate to home screen; press this anywhere and main menu will be displayed.
- **I** — Toggle information view when watching photograph in full-screen mode.
- **1, 2, 3, 4** — Change video playback aspect ratio.
- **Arrow keys** — Navigate menus.
- **Enter** — Select current menu item.
- **Backspace** — Navigate to previous screen.

Social Network Users

This market segment is sometimes referred to by marketers as Generation Y (Gen Y) (this is a succession from the term Generation X, which was made popular by the novel of the same name) and it is generally understood to refer to people born in the early 1980s and 1990s.

This is evidently a large market and is consequently of considerable interest to device manufacturers and ISVs.

People in this segment are comfortable with the Internet, particularly within social networks such as Facebook, Twitter, and MySpace. A survey of 7,705 college students in the United States, undertaken by Reynol Junco and Jeanna Mastrodicasa for their book *Connecting to the Net.Generation: What higher education professionals need to know about today's students)*, has shown that a key motivator is peer recognition. Other interesting points from this survey are that:

- 97% own a computer.
- 97% have downloaded music and other media using peer-to-peer file sharing.
- 94% own a mobile phone.
- 76% use IM and social networking sites.
- 75% of college students have a Facebook account.
- 60% own some type of portable music and/or video device such as an iPod.
- 49% regularly download music and other media using peer-to-peer file sharing.
- 34% use websites as their primary source of news.
- 28% author a blog and 44% read blogs.
- 15% of IM users are logged on 24 hours a day, 7 days a week.

Consequently, areas considered important for this market segment are branding, content, and device design.

A Practical Example

An interesting example of a complete operating system targeting this niche is gOS 3 Gadgets, found at `http://thinkgos.com/index.php`. gOS 3 Gadgets is a full desktop operating system for desktops and notebooks.

This is based on Ubuntu 8.04 and it launches Google Gadgets for Linux on startup, introducing over 100,000 possible iGoogle and Google Gadgets to the desktop. Google Gadgets are lightweight XML programs that allow you to place interactive data inside widgets, and iGoogle is a customizable AJAX-based start page to access these widgets.

Chapter 6: Application Selection

AJAX is a set of programming techniques that enable you to seamlessly update a web page or a section of a web application with input from the server, but without the need for an immediate page refresh.

gOS applications that launch in Mozilla Prism windows closely resemble the functionality of a desktop application. Mozilla Prism also supports WINE 1.0, giving users access to Windows software.

Mozilla Prism enables applications to run in their own window rather than in a browser. It allows a developer access to operating system features that are common to most desktop applications.

Google Gadgets can be installed using apt:

```
$ sudo apt-get install google-gadgets-gtk
```

There is also a QT version available in the repositories google-gadgets-qt.

To run the application, open a terminal and run:

```
$ ggl-gtk
```

To start the application at system boot on Ubuntu Netbook Remix, go to Preferences ➪ Startup Applications and click Add. In the name field, type **Gadgets** and in the command line, put the following:

```
ggl-gtk
```

Now each time the device starts, Google Gadgets will autostart and present the desktop shown in Figure 6-4.

Figure 6-4

135

Chapter 6: Application Selection

A good way to learn about the gadget API is to modify an existing gadget. The following sections will show how to create a modified hello-world gadget.

Set Up the Environment

Make sure that Google Desktop (http://desktop.google.com) and the latest version of SDK (http://desktop.google.com/downloadsdksubmit) are installed.

Copy the Gadget

Copy the api/samples/gadgets/HelloWorld folder and name it **UbuntuMobile**.

Modify It

Inside the folder are the strings that are used in the application. Change the file to look like this:

```
<strings>
<GADGET_NAME>Professional Ubuntu Mobile Development Gadget</GADGET_NAME>
<GADGET_DESCRIPTION>Demonstrates how to write a Gadget</GADGET_DESCRIPTION>
<GADGET_COPYRIGHT>Copyright (c) 2007 Google Inc.
All Rights Reserved</GADGET_COPYRIGHT>
<GADGET_WEBSITE>http://desktop.google.com/</GADGET_WEBSITE>
<GADGET_ABOUT>Professional Ubuntu Mobile Development Gadget
Original Copyright (c) 2007 Google Inc. All Rights Reserved
Modified by Ian Lawrence under the Apache License, Version 2.0
http://www.apache.org/licenses/LICENSE-2.0 for the book

Professional Ubuntu Mobile Development
This gadget displays the text 'Professional Ubuntu Mobile Development'
on top of a picture using the main.xml design file.
If the text is clicked on, a message box will show up as
 an event.</GADGET_ABOUT>
<PUMD>Professional UME Dev</PUMD>
</strings>
```

Change main.xml to look like this:

```
<view width="250" height="150">
  <script src="main.js" />
  <img src="background.png" />
  <label x="125" y="70" align="center" width="250" size="15" enabled="true"
    name="mylabel" pinX="125" pinY="8"
    onclick="onTextClick();">&PUMD;</label>
</view>
```

Add animation into main.js:

```
function onTextClick() {
  view.beginAnimation(doRotation, 0, 360, 500);
}
function doRotation() {
  mylabel.rotation = event.value;
}
```

Chapter 6: Application Selection

Now it is possible to run the application.

Double-click on gadget.gmanifest and install using the gadgets application that was installed earlier. This will present the install message shown in Figure 6-5.

Figure 6-5

You'll also see the gadget with a nice animation, which is shown in Figure 6-6.

Figure 6-6

Also update the metadata in gadget.manifest to reflect the changes in strings.xml.

To distribute the gadget, create a Zip file (right-click on the UbuntuMobile folder and choose Create Archive), and then change the extension from .zip to .gg and users can click the .gg file to install the application to their netbook.

Location-Aware Users

Users with aging parents and young children find the concept of location important to them. This user group also has a relatively high disposable income as they have spent some years working and so present an attractive market to an OEM. Features that users generally expect from a location-aware mobile device include the following:

- The capability to automatically reconfigure itself, such that it always uses the right network settings (for firewall proxies and VPN), and prints on the right printer (i.e., use the home printer when at home, the work printer when at work)
- Security, allowing access only from designated physical locations
- Help to find it if it is misplaced, or stolen
- The ability to easily share a paper or presentation with everyone else in a room, without having to e-mail it or set up a folder on a common server

Some of this functionality could be presented to the user using a simple web interface running locally on the device.

A Practical Example

In this section, you learn how to create a GPS-enabled standalone web application for a MID device. It embeds a web server and database inside a Python application, which receives information from the D-Bus interface using XML-RPC. It uses the Django web application framework.

D-Bus is covered in more detail in Chapter 4.

Background

Traditionally, GPS was provided on Debian-like systems using a daemon called gpsd (`apt-get install gpsd`). There are, however, some problems using gpsd in resource-constrained environments, most notably that gpsd is designed to be run at system start, and stopped when the system is shut down (or a USB/Bluetooth hotplug device started when that device is plugged-in or removed). Therefore, if a client is started before gpsd is running (for example, a GPS applet is started before the Bluetooth device is connected), then the only way for it to know about gpsd is for it to attempt to connect to the gpsd socket and keep trying every so often until it succeeds.

This makes gpsd client programs very busy, always having to wake up to check to see if they can connect. On a system that runs on batteries, having processes that can't sleep very often is a bad thing. Also there is a problem with the granularity of the info that gpsd provides, as running clients get notified about everything, even if they don't care about it. GPS emits a new fix on every NMEA sentence received, which for most GPS devices is about five a second, and each time the clients are all woken up even if they don't care about the data that has changed.

gpsd has no way for clients to say that they are only interested in position data, or only in whether the device has a fix or not. Even if the GPS unit is stationary, satellite data is constantly changing, and the clients will be woken up on every message. Because of these problems Iain Holmes at o-hand wrote gypsy. This is a GPS multiplexing daemon that gives finer control of GPS info and allows programs to call the following objects:

- **Gypsy Position** — For location information
- **Gypsy Course** — For course information
- **Gypsy Accuracy** — For accuracy information
- **Gypsy Satellite** — For satellite information

Implementation

First, install Django:

```
$ sudo apt-get install python-django
```

In the Home directory, create a directory called "Web" and move into it like this:

```
$ mkdir Web;cd Web
```

Next, create a directory:

```
$ mkdir django_projects
```

Move into the Django projects directory and start a new project called "locate" using Django's command line utility. This will create a basic directory structure and the necessary configuration files:

```
$ cd ~/Web/django_projects
$ django-admin.py startproject locate
```

Do not set up apache, postgresql, or mysql. (These applications are not suitable for an embedded device.) The application will use the Cherry Py standalone WSGI server and SQLite.

Edit settings.py so that it looks like this:

```
DATABASE_ENGINE = 'sqlite3'
DATABASE_NAME = os.path.join(os.path.dirname(__file__), 'data', 'locate.db')
DATABASE_USER = ''          # Not used with sqlite3.
DATABASE_PASSWORD = ''      # Not used with sqlite3.
```

The code `DATABASE_NAME = os.path.join(os.path.dirname(__file__), 'data', 'locate.db')` is a trick that enables the SQLite filesystem database to be stored in the project inside the folder

Chapter 6: Application Selection

called data. This is an advantage when using revision control systems in that all code can be stored in a branch. A similar trick works for the site media — create a folder called media in the project root and then in settings.py add the following:

```
import os,sys
path = os.path.dirname(sys.argv[0])
full_path = os.path.abspath(os.path.join(path, '../media'))
```

Then set a `STATIC_ROOT` variable equal to full_path:

```
STATIC_ROOT = full_path
```

Finally, in urls.py import the Django settings from django.conf. Set the static.serve to

```
(r'^media/(?P<path>.*)$', 'django.views.static.serve',
{'document_root': settings.STATIC_ROOT, 'show_indexes': True}),
```

Also create a folder in the project root called "server." Inside this folder, check out a copy of the standalone server:

```
$ wget http://svn.cherrypy.org/trunk/cherrypy/wsgiserver/__init__.py -O wsgiserver.py
```

Create a file called run.py with the following in it:

```
import wsgiserver
#This can be from cherrypy import wsgiserver if you're not running it standalone.
import os
import django.core.handlers.wsgi

if __name__ == "__main__":
    os.environ['DJANGO_SETTINGS_MODULE'] = 'locate.settings'
    server = wsgiserver.CherryPyWSGIServer(
        ('127.0.0.1', 8000),
        django.core.handlers.wsgi.WSGIHandler(),
        server_name='Lowkate',
        numthreads = 20,
    )
    try:
        server.start()
    except KeyboardInterrupt:
        server.stop()
```

Create a desktop file in /usr/share/applications that will call a script to start the server and the midbrowser (a customized version of Firefox).

Chapter 6: Application Selection

```
[Desktop Entry]
Encoding=UTF-8
Version=1.0
Type=Application
Name=Locate
Comment=Professional Ubuntu Mobile Development
Icon=locate
Categories=Application;Utility;
Exec=/usr/bin/locateit
```

Also place an image in /usr/share/icons/hicolor/48x48/apps. The locateit script looks like this:

```
export PYTHONPATH="/home/ubuntu/Web/django_projects/locate/:$PYTHONPATH"
export PYTHONPATH="/home/ubuntu/Web/django_projects/:$PYTHONPATH"
export DJANGO_SETTINGS_MODULE=locate.settings
python /home/ubuntu/Web/django_projects/locate/server/run.py&
midbrowser http://127.0.0.1:8000/geoapp/
```

Clicking the icon shown in Figure 6-7 executes the script and starts the geo app.

Figure 6-7

The locate project should now look like this:

```
$ ls
data media server __init__.py manage.py settings.py urls.py
```

It is a completely self-contained project environment.

141

Chapter 6: Application Selection

Test the Gypsy to GPS Connection

Install the gypsy daemon and library by adding the following:

```
deb http://ppa.launchpad.net/ianlawrence/ppa/ubuntu jaunty main
deb-src http://ppa.launchpad.net/ianlawrence/ppa/ubuntu jaunty main
```

to /etc/apt/sources.list and then do the following:

```
$ sudo apt-get update && sudo apt-get install gypsy
```

The .debs (for 5.1) are also available at http://ianlawrence.info/downloads/debian/gypsy-lpia/gypsy_0.5-1_lpia.deb *and* http://ianlawrence.info/downloads/debian/gypsy-lpia/libgypsy0_0.5-1_lpia.deb.

The bluetooth GPS device that was used was the Nokia Wireless GPS Module LD-3W. To communicate with it, you need to find out its address. Make sure bluetooth is enabled:

```
$ sudo apt-get install bluetooth bluez-utils bluez-gnome gnome-bluetooth
libbluetooth2 libbtctl4 libgnomebt0 nautilus-sendto
```

Run the following:

```
user@host:~$ hcitool scan
Scanning ...

00:19:B7:8C:A7:F7 IansPhone
00:02:76:C5:58:B2 Nokia LD-3W
```

This returns the address of the GPS device: 00:02:76:C5:58:B2.

Interaction with the GPS Daemon

The code to interact with the gypsy daemon (gypsy.py) and to show its status (status.py) was written by Ross Burton and released LGPL and GPL respectively. It can be checked out by running the following:

```
$ bzr branch http://burtonini.com/bzr/gypsy-status/
```

Add the bluetooth address into gypsy.py:

```
dbus.mainloop.glib.DBusGMainLoop(set_as_default=True)
gps = GPS("00:02:76:C5:58:B2")
gps.Start()
```

Then, inside a target terminal, add the following:

```
export DISPLAY=:2
python status.py
```

Chapter 6: Application Selection

The status of the GPS is displayed, as shown in Figure 6-8, along with a small map of the location pulled from a mapping API.

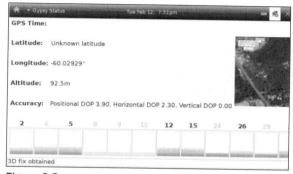

Figure 6-8

D-Bus and HTTP Requests

It is not possible to query D-Bus without starting some sort of a loop — such as the call to `DBusGMainLoop`.

You can see it in status.py. This loop condition will not happen in a locally running web application, so some alternative solution is necessary.

XML-RPC is a way of allowing software running on disparate operating systems and running in different environments to make procedure calls over the Internet. Its remote procedure calls using HTTP as the transport and XML as the encoding. Django has an XML-RPC server already available, so install the code that you'll find at http://code.google.com/p/django-xmlrpc/ inside the application. Make a method in settings.py like this:

```
XMLRPC_METHODS = (
# List methods to be exposed in the form (, ,)
('locate.django_xmlrpc.views.handle_xmlrpc', 'handle_xmlrpc',),
('locate.django_xmlrpc.views.handle_gypsy', 'handle_gypsy',),
```

In django_xmlrc/views.py, create a method signature like this:

```
@xmlrpc_func(returns='string', args=['string', 'int', 'int', 'int',])
def handle_gypsy(timestamp, latitude, longitude, altitude):
    """Take the values we need from the the gypsy XML object and store them in a
    database (or file system) """
    data = Raw(time=timestamp, lat=latitude, long=longitude, alt=altitude)
    data.save()
    # This returns the values passed in (useful for debugging)
    return "The timestamp is %s, the latitude is %i, the longitude is %i, and the
altitude is %i." % (timestamp, latitude, longitude, altitude)
```

The XML-RPC client needs to pass the GPS XML object to the `handle_gypsy` function that is shown in the Django Administration dialog box in Figure 6-9.

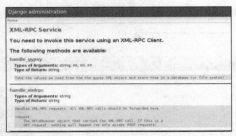

Figure 6-9

Also create a models.py file inside the django_xmlrpc application to store the GPS data from django.db import models:

```
class Raw(models.Model):
    time = models.CharField(max_length=200)
    lat = models.FloatField()
    long = models.FloatField()
    alt = models.FloatField()

def __unicode__(self):
        return self.time
```

In gypsy.py, add the following:

```
import xmlrpclib
```

Next, add the following:

```
    gps.Start()

    server = xmlrpclib.ServerProxy('http://127.0.0.1:8000/xmlrpc/')

    def position_changed(fields, timestamp, latitude, longitude, altitude):
        server.handle_gypsy(timestamp, latitude, longitude, altitude)
```

The code passes the GPS data into the Django web application.

All that remains to do is to query the database for the last GPS position so that it can be used in the web interface. Create an application called geoapp with a views.py file like this:

```
# Create your views here.

from django.http import HttpResponse
from locate.django_xmlrpc.models import Raw
from django.shortcuts import render_to_response
```

```python
def get_position(request):
    latest_GPS = Raw.objects.all().order_by('-id')[:1]

    return render_to_response('location_list.html', {'latest_GPS': latest_GPS})
```

Next, create a template:

```
<html>
<head></head>
<body>
{% if latest_GPS %}
   <ul>
   {% for co in latest_GPS %}
    <li>
     I am located at <a href="http://geohash.org/?q={{co.lat}},{{co.long}}">this</a> URL.</li>
   {% endfor %}
   </ul>
{% else %}
   No coordinates
{% endif %}
</body>
</html>
```

This uses geohash.org, which is a latitude/longitude geocode system invented by Gustavo Niemeyer when writing the web service at geohash.org, and put into the public domain. It offers short URLs to uniquely identify positions on the Earth, so that referencing them in e-mails, forums, and websites is more convenient.

Figure 6-10 shows where we were when writing this code (and also when writing this book).

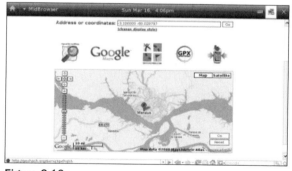

Figure 6-10

Chapter 6: Application Selection

Summary

This chapter highlighted some of the potential groups of users that an OEM can target when thinking about the "look and feel" of a device. This is not an inclusive list, and there are other important groups. For example, one type of user that was identified by members of the Ubuntu Mobile team was in Asia, particularly in Japan, where many people use a device solely for reading e-books while taking public transportation. Determining how to target such groups is left to you. Such a challenge requires an OEM to think creatively about how to differentiate its offering from others on the market. We hope this chapter has provided some insight into this process.

7
Theming

Theming allows a user or device manufacturer to customize the look and, to some extent, the feel of the operating system, window manager, or widget set.

This chapter begins by showing the theme structure on Ubuntu MID and Ubuntu Netbook Remix through the gtkrc file and GConf. This is followed by a fully customized Ubuntu MID theme. Finally, theme performance and optimization are covered along with some suggestions for further investigation.

What Is a Theme?

A GTK+ theme consists of three things:

- A theme engine (sapwood in the case of Ubuntu MID), which is a shared object and includes code to draw the graphical elements
- A configuration file to set the operation of the theme engine and the core parts of GTK+2
- Extra data files for the theme engine, such as images

Not included in the context of themes are:

- Icons
- Backgrounds
- Fonts
- Sounds

Not all themes have extra images and data.

Chapter 7: Theming

On Ubuntu MID, these elements are usually installed or configured by the distributor or OEM per their preferences, and there are currently no packages and management tools available to integrate these elements into a theme.

Where Are Themes Located in the Filesystem?

Themes are located in `/usr/share/themes` or in the user's home .themes directory.

What Is a Theme Engine and Where Are They Located?

GTK engines are built to render buttons and everything that is "inside" a GTK window. That's why they're called GTK engines. There are several GTK engines and they are located in: `/usr/lib/gtk-2.0/2.10.0/engines/`

Of these, the Ubuntu MID software release uses the sapwood engine (libsapwood.so). Others included in the package gtk2-engines include:

- Clearlooks, the default GNOME theme, based on Bluecurve
- Crux, formerly known as the Eazel engine
- High contrast, which is used by some accessibility themes
- Industrial, the famous engine from Novell (formerly Ximian)
- LighthouseBlue, another engine based on Bluecurve
- Metal, which gives a metallic look
- Mist, a flat and high performance engine
- Redmond95, which provides a look similar to that of Windows
- ThinIce

Theming Ubuntu MID

Sapwood provides a server and client library for accessing theme images. The server is responsible for loading the theme-related images and distributing them to clients. Sapwood saves memory compared to, for example, the Pixbuf engine as it shares images between applications using a cache.

Sapwood is also much faster because it tiles the 16-bit images using the X Server.

It is possible to observe this in the following location:

```
$ /usr/share/ubuntu-mid-default-settings/mid-gui-start
```

After first checking that Compiz is not enabled, the script tells the Matchbox Window Manager to use the mobilebasic theme:

Chapter 7: Theming

```
/usr/bin/matchbox-window-manager \
    -theme /usr/share/themes/mobilebasic/matchbox/theme.xml \
```

Compiz is a compositing window manager for the X Window System, which uses 3D Graphics hardware to create desktop effects for window management.

What Happens When a MID Device Boots?

On boot, upstart will execute `/etc/event.d/session`, which will launch the X Window System using the following command:

```
exec openvt -e -s - su -l ubuntu -c "env -u XDG_SESSION_COOKIE startx -- -br"
```

Once X starts, it parses the session scripts in `/etc/X11/Xsession.d/`, including the `/etc/X11/Xsession.d/25midstartup file`, which has a call

```
MID_GUI=/usr/share/ubuntu-mid-default-settings/mid-gui-start
```

On older versions of Ubuntu Mobile there was an export of the `GTK2_RC_FILES` environment variables for GTK+2 widget themes in this file, as in the following:

```
export GTK2_RC_FILES=/usr/share/themes/mobilebasic/gtk-
2.0/gtkrc:/usr/share/themes/mobilebasic/gtk-2.0/gtkrc.maemo_af_desktop
```

These files define all the GTK+2 styles and bindings that will create the look and feel of all the widgets in the applications and in the Hildon desktop.

On Ubuntu Jaunty, the GTK theme changes are now made by setting the following GConf key to the name of a theme directory: `/desktop/gnome/interface/gtk_theme`. For example, this could be set to `mobiletheme` or any other valid theme in the theme directory by using the following:

```
gconftool-2 --set --type string /desktop/gnome/interface/gtk_theme mobiletheme
```

When that key changes, the change is propagated to xsettings and, through this, all GTK apps are notified of the new theme, and they immediately redraw themselves. GTK applications launched later than the key change will automatically start using the new theme.

GConf, which was discussed in depth in Chapter 4, is a system for storing application preferences. It exists so that application preferences can be more easily managed. It is also process transparent. This means that a setting changed in one application will instantly update in all other applications that are interested in that setting. This technology is vital for the application of coherent user interfaces on mobile devices.

The easiest way to manage GConf settings is to use `gconf-editor`:

```
$ sudo apt-get install gconf-editor
$ gconf-editor
```

Chapter 7: Theming

This displays an interface according to the settings that are available. In addition to using the front-end gconf-editor, settings can be inspected using `gconftool`. Run the following command:

```
$ gconftool-2 -R /desktop/gnome
```

This displays the Ubuntu Mobile Desktop settings. Run the following command:

```
$ gconftool-2 --get-default-source
```

This shows the location of the XML backend, which GConf uses as a database.

Modifying Themes

It is often necessary to customize themes, particularly when working with an OEM. Before doing this, you might want to read "Visual Design of the GNOME Human Interface Guidelines," at `http://library.gnome.org/devel/hig-book/stable/`, which starts as follows:

> Visual design is not just about making your application look pretty. Good visual design is about communication. A well-designed application will make it easy for the user to understand the information that is being presented, and show them clearly how they can interact with that information.

Substitute the word "application" for "theme" in the preceding quote and you should have a good idea about the objective of theming.

A Useful Tool When Working with Themes

This following bash script displays information on the current theme. It is useful to use when working with themes and GConf.

First, install zenity, which is a handy suite of GTK widgets:

```
$ sudo apt-get install zenity
```

Call the script that follows **get_theme_info.sh** and execute it like `./get_theme_info.sh` (after making it executable):

```
#!/bin/bash
icons=$(gconftool-2 -g /desktop/gnome/interface/icon_theme)
theme_gtk=$(gconftool-2 -g /desktop/gnome/interface/gtk_theme)
theme_metacity=$(gconftool-2 -g /apps/metacity/general/theme)
font_gtk=$(gconftool-2 -g /desktop/gnome/interface/font_name)
font_meta=$(gconftool-2 -g /apps/metacity/general/titlebar_font)
zenity --info --text "
icons: $icons
GTK theme: $theme_gtk
metacity theme: $theme_metacity
gtk font: $font_gtk
metacity font: $font_meta
Professional Ubuntu Mobile Development"
```

Chapter 7: Theming

The result displays the information that's shown in Figure 7-1.

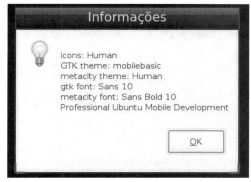

Figure 7-1

Theme Structure

This section explores the mobile theme on Ubuntu MID. This theme is based on the structure of themes in Hildon, which Ubuntu inherited from upstream. The layout (from /usr/share/themes/mobilebasic) looks like:

```
/usr/share/themes/mobiletheme/gtk-20/
      index.theme
      gtkrc
      gtkrc.cache
images/
      *.png
matchbox/
      theme.xml
```

The two main files within this structure are theme.xml and gtkrc.

The theme.xml File

Some simple window decoration customization can be achieved by modifying things within the theme.xml file. It defines things such as the Main title bar decoration and some dialog decorations as well.

The three basic resources in a theme.xml file are:

- ❑ Colors
- ❑ Fonts
- ❑ Images

Before using them, declare them all in the format:

```
id :definition
```

Chapter 7: Theming

For example

```
<type id="object name" definition="data" />
```

where the `type` is the color, font, pixmap, etc, `id` is the object's unique name, and `definition` is the metadata to be set.

A color is defined as follows:

```
<color id="osso-DialogTitleTextColor" def="#ffffff" />
```

A font is defined as follows:

```
<font id="osso-TitleFont" def="Nokia Sans-17.85" />
```

An image is defined as follows:

```
<pixmap id="dialoguptile" filename="../images/qgn_plat_dialog_frame_tile_top.png" />
```

The gtkrc File

More complex customizations involve changing the gtkrc file. This file is common to theming on both Ubuntu Netbook Remix and Ubuntu MID; conceptually, it is useful to think that the GTK theme "lives" inside the gtkrc file. The file format is case-sensitive, comments are marked with a # (hash). Multi-line comments are also possible with comments similar to the C language such as `/* */`.

A gtkrc file is a text file composed of a sequence of declarations. Hash (#) characters delimit comments and the portion of a line after a # is ignored when parsing a gtkrc file.

This file gtkrc is not very tolerant to typos and errors so be careful when changing the code.

Customizing a gtkrc File

The following gtkrc file applies a theme to all widgets (buttons, scrollbars, edit boxes).

Technically, every widget is derived from `GtkWidget`. *This means, changes to the properties of* `GtkWidget` *will affect all widgets and the concept of inheritance applies. Properties of* `GtkButton` *will also be applied to* `GtkCheckButton` *and so on unless it is explicitly stated otherwise. The complete GTK Class hierarchy is available at* `http://library.gnome.org/devel/gtk/stable/ch01.html`.

It does this by specifying a *style*. A style can either be applied to the actual GTK theming system itself such as in the following code or to the specific theme "engine" that's used by the theme (in the case of Ubuntu Netbook Remix, Murrine, and with Ubuntu MID, sapwood) as follows:

```
style "default-style"
{
        # modify the x/y thickness to determine padding between text and border of
        # widgets.
        xthickness = 2
        ythickness = 2

        # style on a slider
        GtkRange::slider-width = 15
```

```
            # set the background to a light grey
            bg[NORMAL] = "#f6f6f6"
            # and the forground to black
            fg[NORMAL] = "#000000"
}

class "GtkWidget" style "default-style"
```

The code creates a style called default-style and then sets some padding on the widgets, styles a slider, sets some colors, and applies the style using the class `GtkWidget`. These are explained in the following sections.

Padding

This sets the padding using an x,y coordinate system.

Styles

Style properties are in the form

```
WidgetName::style-property-name = VALUE
```

The preceding slider looks like this:

```
GtkRange::slider-width = 15
```

The global picture theming style for the mobilebasic theme on MID looks like this:

```
# global picture theming.
   GtkWidget::hildon-focus-handling = 1
   GtkWidget::focus-line-width = 0
   GtkWidget::focus-padding = 0
```

Colors

It is possible to change some sections of the gtkrc file to match our theme's colors. In the preceding code, color takes the following format:

```
bg[NORMAL] = "#f6f6f6"
```

bg stands for background. The following table describes the four valid categories:

Category	Usage
fg	Foreground color. Used for text on buttons. Also used for the button borders in some engines.
bg	Background color. This is the background color of windows and buttons.
text	Text color for text input widgets and lists.
base	Background color of text widgets and lists.

153

Chapter 7: Theming

Applying the Style

In GTK, there are three ways to apply styles — the `class`, `widget_class`, and `widget` statements, of which the `class` and `widget` are the most important.

In the preceding code, the style was applied by this call:

```
class "GtkWidget" style "default-style"
```

When finding styles that are relevant to the current widget, a widget hierarchy applies so that the style will be applied to whichever widget class name matches the one in the code. The hierarchy looks something like this (in order of importance):

- GtkWidget
- GtkContainer
- GtkBin
- GtkButton
- GtkToggleButton

In the preceding example, styles are applied to the top-level widget (GtkWidget) and then cascaded down.

The use of `widget_class` to apply styles is not covered here. `widget` matches work on the names of the widgets. This is very useful if there is a specific widget in an application that needs to be modified.

GTK applies the different "styles" in the preceding order. Styles that merged later on override the settings of the earlier ones. This makes sense, as the `class` matches are very broad while the `widget` matches are very specific and often pick out just a single widget.

Theming Ubuntu MID

You can theme two ways: manually and automatically.

Manually Theming MID

To create a theme from scratch that's based on the existing mobilebasic theme on Ubuntu MID, copy the default mobilebasic theme to a new directory called pumd-theme and then move it into the Matchbox directory.

It is also possible to use `hildon-theme-tools` and the command `hildon-theme-bootstrap` to create and then package a new theme. This is explained in the section "Automatically Theming MID."

```
$ cd /usr/share/themes/
$ sudo mkdir pumdtheme
$ sudo cp -r mobilebasic/* pumdtheme
$ cd pumdtheme/matchbox
```

Chapter 7: Theming

Open the file theme.xml in this folder and change the metadata first line to

```
<theme name="pumdtheme" author="Ian Lawrence" desc="A theme created for the book
Professional Ubuntu Mobile Development, published by Wiley" version="0.1"
engine_version="1" cache="false" >
```

Next, change the following:

```
<color id="osso-TitleTextColor" def="#eec73e" />
<color id="osso-DialogTitleTextColor" def="#403008" /
```

For the new MID theme, change the color definitions in the gtkrc file in the gtk-20 folder to the following:

```
color["DefaultTextColor"]         = "#eec73e"
color["EmpTextColor"]             = "#204a87"
color["PaintedDefaultTextColor"]  = "#ffffff"
color["DisabledTextColor"]        = "#babdb6"
color["ProgressTextColor1"]       = "#000000"
color["ProgressTextColor2"]       = "#000000"
color["SecondaryTextColor"]       = "#5a5a5a"
color["TitleTextColor"]           = "#000000"
color["DialogTitleTextColor"]     = "#403008"
```

These steps change the color of the fonts in the menus.

Now open the file index.theme in the main directory and change it to the following:

```
[Desktop Entry]
Type=X-Hildon-Metatheme
Name=PUMDTheme
Icon=/usr/share/themes/pumdtheme/images/qgn_plat_theme_thumbnail.png
Encoding=UTF-8
[X-Hildon-Metatheme]
GtkTheme=pumdtheme
IconTheme=gnome
X-MatchboxTheme=pumdtheme
X-OperaSkin=default.zip
```

Also modify the appropriate images in the images folder. A tool that's useful when working with images is called xmag. It is provided by the package x11-apps:

```
apt-get install x11-apps
```

x11-apps provides a miscellaneous assortment of X applications that ships with the X Window System, including xmag, x11perf, and x11perfcomp, which are tools for benchmarking graphical operations under the X Window System (see the section "Performance Testing of Themes").

Making all the changes above and changing the appropriate images will create a new theme. For example, the theme that's shown in Figure 7-2 is based on the Intrepid release of Ubuntu MID. If you do not have new images to add it is possible to clearly see the changes made by opening a terminal (the terminal color changes).

155

Chapter 7: Theming

Figure 7-2

It is possible to include the .gtkrc-2.0 file in the default user home directory that loads the theme (this works because it loads the XSETTINGS for the icons).

> *XSETTINGS is a mechanism on Linux that allows the configuration of settings and then allows these settings to be propagated across all applications at runtime.*

A simple gtkrc-2.0 file in a user's home directory can just be an include file, as shown below:

```
$ cat ~/.gtkrc-2.0
include "/usr/share/themes/pumdtheme/gtk-2.0/gtkrc"
```

Automatically Theming MID

In the previous section, we manually themed Ubuntu MID. It is possible to automate this process through the use of the hildon-theme-tools package. This is installed using apt:

```
$ sudo apt-get install hildon-theme-tools
```

The package contains various tools that can be used for theming on Ubuntu MID. Begin by installing the subversion package

```
$ sudo apt-get install subversion
```

and then run the following:

```
$ hildon-theme-boostrap
```

This will download a skeleton theme into the specified directory. Move into the template directory and edit the template.svg as required using an image editor such as GIMP.

> *Make sure you select the second option, 2) hildon-theme-layout-4, since layout-3 is not supported by MID.*

Next, to build a debian package of the theme, it is necessary to change a couple of things in the debian folder. In the control file, remove the dependency on the hildon-theme-cacher. In the postinst file, remove the line above DEBHELPER, which calls the theme cacher utility:

> *hildon-theme-cacher caches the gtkrc files — it is not used on Ubuntu MID.*

Next, install the dependencies that are required to build the theme package:

```
$ sudo apt-get install fakeroot hildon-theme-layout-4 cdbs debhelper dpkg-dev
```

Build the package by running the following:

```
$ dpkg-buildpackage -tc -rfakeroot
```

This will create both a .tar.gz and the .deb — for example, hildon-theme-pumd_4.2.0-1_all.deb.

The Makefile.am in the data directory calls hildon-theme-subst (another tool in the hildon-theme-tools package), which makes substitutions in gtkrc files.

The resulting .deb can then be uploaded to a PPA or installed locally using the following:

```
$ dpkg -i hildon-theme-pumd_4.2.0-1_all.deb
```

Theming Ubuntu Netbook Remix

Theming Ubuntu Netbook Remix is done in the same way as theming the main Ubuntu distribution itself.

Actually, the theme that is referred to in this section is an XFCE theme, but because XFCE uses the Murrine engine, it also works nicely on UNR.

The theme itself is based on an existing theme called MurrinaSegPhault v0.02, which is itself composed of two engines — Murrine and Clearlooks. This is a nice innovative technique; it uses the Clearlooks rendering engine for specific widgets and Murrine for everything else (a big nod to Ryan Paul at arstechnica for this discovery).

In order for this to work, the engine that will override the default needs to be specified inside a style block in a gtkrc file like this:

```
style "theme-notebook" = "theme-wide"
{
  base[SELECTED]   = "#e3bd3b" # Tab selection color
  bg[ACTIVE]       = "#e4d79e" #  Unselected tabs
  engine "clearlooks" {
    style = GLOSSY
  }
}
```

The "engine" keyword is used to specify the desired theme engine for rendering the specific element, and the braces that follow can be used to set specific options for that theme engine.

The creation of the actual package, xfce4-theme-ubuntu-golden.deb, is covered in Chapter 5.

To see the actual gtkrc file for the theme, add

```
deb http://ppa.launchpad.net/ianlawrence/ppa/ubuntu jaunty main
deb-src http://ppa.launchpad.net/ianlawrence/ppa/ubuntu jaunty main
```

Chapter 7: Theming

to `/etc/apt/sources.list` and then do

```
sudo apt-get update && apt-get source xfce4-theme-ubuntu-golden
```

Other parts of the UNR system can be customized, as explained below.

Boot Splash

Usplash is an application that draws a splash screen at boot with a throbber to show the boot process.

To customize this, download the source of a usplash theme (here usplash-theme-ubuntu) using the following code:

```
$ sudo apt-get source usplash-theme-ubuntu
```

and move into the folder and customize the images. Optionally, create new images (with a maximum depth of 256 colors) and also customize usplash-theme-ubuntu.c. When the changes have been made, run the following:

```
$ sudo dpkg-buildpackage -d
```

which will create the new usplash package.

Karmic usplash will only be used in situations where the boot process needs to be delayed for some reason – for example, to fsck a filesystem or to ask for an encryption password. usplash is being replaced by xsplash (`https://edge.launchpad.net/xsplash`), which is a new tool written by Canonical to bring the X Server up as quickly as possible to help speed up the boot process.

Creating a gdm Theme

A custom gdm theme can be made for UNR — an example is shown in Figure 7-3.

Figure 7-3

This theme consists of some images — the background, a screenshot, and some icons — and a GdmGreeterTheme.desktop file, along with an XML file, which contains information about the theme layout.

Chapter 7: Theming

A good way to learn how to theme the Gnome Display Manager is to get the source of an existing theme in the repositories — for example, `apt-get source ubuntustudio-gdm-theme`.

A good way to test a gdm theme during its creation is to make a .tar.gz of all the files mentioned and then to install it to System ⇨ Administration ⇨ Login Window and select the Local tab. Click Add and select the .tar.gz file (the file chooser shows the roots home folder by default so browse to the directory where the tar.gz was saved) and then select the radio button next to the theme. After doing this, make sure xnest is installed (`apt-get install xnest`) and then run the following:

```
$ gdmflexiserver --xnest
```

The new gdm theme will be displayed in a window. This method can also be used to take the screenshot for the package itself. When finished, roll a new debian package using `dpkg-buildpackage`:

```
$ sudo dpkg-buildpackage -d
```

Customizing the Netbook Launcher

It is possible to customize the launcher on Ubuntu Netbook Remix. The files are located in `/usr/share/netbook-launcher`.

When installed on UNR, xfce4-theme-ubuntu-golden looks like Figure 7-4.

Figure 7-4

After the netbook-launcher is customized, the device looks like Figure 7-5.

Figure 7-5

Ubuntu Netbook Remix is now customized from the initial boot splash to the desktop.

Chapter 7: Theming

Performance Testing of Themes

It is possible to run a range of performance tests on themes. The simplest test to run is one using the metacity-theme-viewer.

> *Metacity is a compositing window manager that is used by default in GNOME.*

Run the following command on UNR:

```
$ metacity-theme-viewer Human
```

It outputs the following results:

```
Loaded theme "Human" in 0.02 seconds
Drew 100 frames in 0.61 client-side seconds (16.1 milliseconds per frame) and
3.4817 seconds wall clock time including X server resources (34.817 milliseconds
per frame)
```

Test the Human Metacity Theme

The theme should render properly in all the tabs. There's also a Benchmark tab that can be used to compare rendering speeds with other themes. For example, the Human theme on Ubuntu Netbook Remix takes 34.817 milliseconds to draw one window frame, whereas the Simple theme takes 8.84145 milliseconds.

Another benchmarking tool for GTK+ themes is called GtkPerf. It is available through apt (sudo apt-get install gtkperf). This tool works also on Ubuntu MID.

It can be run like this:

```
gtkperf -a -c 1000
```

Comparisons

Here are the results of the test for the theme and netbook launcher in the previous section. It is for Ubuntu Netbook Remix running on an LPIA base. The device is an Acer Aspire One:

```
GtkPerf 0.40 - Starting testing: Sun Jun  7 21:00:07 2009
GtkEntry - time:  1.31
GtkComboBox - time: 53.03
GtkComboBoxEntry - time: 35.79
GtkSpinButton - time:  7.05
GtkProgressBar - time: 15.15
GtkToggleButton - time:  9.06
GtkCheckButton - time:  7.59
GtkRadioButton - time: 11.90
GtkTextView - Add text - time: 170.83
GtkTextView - Scroll - time: 17.47
GtkDrawingArea - Lines - time: 25.75
```

```
GtkDrawingArea - Circles - time: 34.02
GtkDrawingArea - Text    - time: 31.43
GtkDrawingArea - Pixbufs - time:  3.89
---
Total time: 424.30
```

The following is the result of the same test on the same xfce-ubuntu- golden theme and customized launcher but with an XFCE base:

```
GtkPerf 0.40 - Starting testing: Sun Jun 14 11:36:52 2009
GtkEntry - time:  1.16
GtkComboBox - time: 27.28
GtkComboBoxEntry - time: 20.25
GtkSpinButton - time:  2.94
GtkProgressBar - time:  2.19
GtkToggleButton - time:  2.77
GtkCheckButton - time:  2.25
GtkRadioButton - time:  3.02
GtkTextView - Add text - time: 160.87
GtkTextView - Scroll - time: 12.03
GtkDrawingArea - Lines   - time: 32.52
GtkDrawingArea - Circles - time: 33.27
GtkDrawingArea - Text    - time: 39.06
GtkDrawingArea - Pixbufs - time:  5.12
---
Total time: 344.76
```

The following is the out-of-the-box standard Ubuntu Netbook Remix i386, default theme, and launcher:

```
GtkPerf 0.40 - Starting testing: Sun Jun 14 22:36:05 2009
GtkEntry - time:  1,09
GtkComboBox - time: 31,86
GtkComboBoxEntry - time: 17,61
GtkSpinButton - time:  5,84
GtkProgressBar - time: 16,18
GtkToggleButton - time:  4,71
GtkCheckButton - time:  4,38
GtkRadioButton - time:  8,00
GtkTextView - Add text - time: 152,51
GtkTextView - Scroll - time: 12,92
GtkDrawingArea - Lines   - time: 31,97
GtkDrawingArea - Circles - time: 29,94
GtkDrawingArea - Text    - time: 22,76
GtkDrawingArea - Pixbufs - time:  4,73
---
Total time: 344,51
```

The most obvious difference to note from these results is that the performance on an LPIA base is much worse than on the default Ubuntu Netbook Remix base. Also, there is very little difference between the default netbook remix and the XFCE version — only really in GtkTextView.

Chapter 7: Theming

Setting `Option "RenderAccel" "true"` in the xorg configuration gives a performance boost, particularly with GtkTextView. To do this, back up the current `/etc/X11/xorg.conf`. A good way to do this is to run the following:

```
$ sudo dpkg-reconfigure xserver-xorg
```

Accept the defaults. Make the device section of the new xorg.conf look like this:

```
Section "Device"
      Identifier        "Configured Video Device"
      Option            "UseFBDev"                "true"
      Option            "RenderAccel"             "true"
EndSection
```

> *Setting RenderAccel to true can give some unexpected results such as screen lockups (it's for this reason that it is not enabled by default by the Ubuntu developers). Test this on a device thoroughly.*

Setting this on the XFCE base running Ubuntu Netbook Remix and restarting gave a GtkTextView time of 150.95, a 10 second improvement on the original.

It is also possible to get some performance improvements by using the latest version of the Murrine engine from the GNOME repositories. Murrine Engine is a "shiny" and popular GTK2 engine.

To get this benefit, compile the theme engine itself inside a MID image by adding the multiverse and universe repositories to sources.list, and install the dependencies:

```
$ sudo apt-get install linux-headers-`uname -r` build-essential checkinstall
libgtk2.0-dev imagemagick libtool
```

Download the latest svn:

```
$ svn co http://svn.gnome.org/svn/murrine/trunk/
```

Move inside the source code directory and run the following:

```
$ ./autogen.sh
$ ./configure --prefix=/usr --enable-animation
$ make
$ sudo checkinstall
```

This will pop up some questions. Answer yes to create the default docs and choose a new package description. Then choose a new package version number for the debian policy, and then Enter. This will create a debian file which can be installed using dpkg -i.

X Window Testing

If there is an expensive routine in your theme that cannot be explained through some theme modification, it is possible to test the relative performance of the X Window graphics adapter itself to see if the problem lies there. This is done through the use of x11perf, which attempts to run through most of the X drawing operations and then characterizes how many of these operations the X Server can perform in a given period of time.

Chapter 7: Theming

To show this working, tests were run on an eeePC and Acer Aspire One using the following (this can also be run on MID):

```
$ x11perf -all
```

Both netbook operating systems were fully up-to-date. They were then compared using x11perfcomp, which merges the output of several runs into a nice tabular format.

Overall, the Acer Aspire One performed much better, as you can see in the following results, which is a cherry pick of the operations in which the Aspire One was *much* quicker than the Eee PC. This is shown by the figure in parentheses. For example, on the operation `Fill 100x100 stippled trapezoid (17x15 stipple)`, the Aspire One in column 1 was 2.73 times quicker than the Eee PC in column 2.

```
      1              2          Operation
 --------    ---------------    -----------------
    897.0    2450.0  ( 2.73)    Fill 100x100 stippled trapezoid (17x15 stipple)
   1060.0    3230.0  ( 3.05)    Fill 100x100 opaque stippled trapezoid (17x15 stipple)
    171.0     452.0  ( 2.64)    Fill 300x300 opaque stippled trapezoid (17x15 stipple)
   1510.0    3870.0  ( 2.56)    Fill 100x100 stippled trapezoid (161x145 stipple)
    301.0     552.0  ( 1.83)    Fill 300x300 stippled trapezoid (161x145 stipple)
   2320.0    4920.0  ( 2.12)    Fill 100x100 equivalent complex polygons
```

Summary

A good way to learn theming is to examine other people's themes. This chapter has shown the make-up of a theme and how theming works on both Ubuntu MID and on Ubuntu Netbook Remix. It also discussed basic theme customization and how this can be performance tested. Part of the appeal of Ubuntu for an OEM is that it is fully customizable and that a device can be configured to look exactly as the manufacturer wishes. This is a powerful feature as more and more device manufactures become involved in the mobile device market.

Kernel Fine-Tuning

The kernel is the portion of software that's closest to the hardware. It is responsible for the device management and the system's basic functionality as well as ensuring that the system will be secure and robust enough to run all the applications.

The Linux kernel is fully configurable. It was designed to support a large number of different platforms and for each platform can be set up in many different ways. Ubuntu Mobile has some stable and generic kernel configurations, which support various platforms with a few kernel packages. A consequence of this is that probably an end user will not have to do any extra work to get Ubuntu Mobile running properly on supported hardware. However, this generalization can mean decreased system performance, which is one area of specific interest to OEMs and device manufacturers.

This chapter demonstrates kernel-tuning methods that enable you to create kernel packages. The results are more specific to the target hardware and more optimized for the hardware's usage patterns.

Ubuntu MID Kernel Overview

The Ubuntu MID kernel is based on the LPIA architecture. This architecture was designed to support i386 devices, but uses a different set of configurations from all previous i386-based Ubuntu kernels. LPIA's build options are mainly focused on power management.

Kernel-Tuning Methods

The user can approach kernel optimization using one of two methods.

The first method creates an Ubuntu package. This is a better choice in a situation that has at least one of the following requirements:

Chapter 8: Kernel Fine-Tuning

- A standard Ubuntu kernel package needs to be created.
- The latest development Ubuntu kernel version needs to be used.
- The optimization should be based on an existing Ubuntu kernel package flavor.
- The generated package should be integrated with Ubuntu as much as possible.

The second method results in a Debian package. This is a better choice in a situation that has at least one of the following requirements:

- A Debian package is wanted or at least is enough to start with.
- It is not a requirement to use an Ubuntu kernel.
- The developer thinks it's a simpler method.

If both methods are applicable, then the first one would be a better solution. Although it seems obvious, it's important to mention here that an OEM or user should know the hardware specifications in order to create the customized kernel package capable of running correctly on a target device.

Create an Ubuntu Package

As mentioned earlier, this method creates an Ubuntu kernel package. The kernel source from Ubuntu has plenty of scripts capable of generating various kernel packages in an automated way. Consequently, as a prerequisite to this method, the user should get the source from Ubuntu.

There are two choices when downloading the kernel source. You can obtain it from either a deb or a git repository. In order to have the latest stable Ubuntu kernel source, you have to get it from the Ubuntu git repository. In general, the source from a deb repository won't be as up-to-date as a git repository. So, first we demonstrate the method of getting the source from git and then we cover the use of a deb repository. However, before starting the process, it is necessary to install some dependencies:

```
$ sudo apt-get update
$ sudo apt-get install git-core fakeroot makedumpfile kernel-wedge
$ sudo apt-get install libncurses5-dev build-essential
$ sudo apt-get build-dep linux
```

Now, clone the Ubuntu kernel tree:

```
$ git clone git://kernel.ubuntu.com/ubuntu/ubuntu-jaunty.git \
    ubuntu-jaunty.git
$ cd ubuntu-jaunty.git/
```

The first step is to add the new flavor to the build scripts. Add it to the lpia architecture in both files shown here:

Chapter 8: Kernel Fine-Tuning

```
$ cd debian/
$ cd scripts/misc
$ sed -e s/getall\ lpia\ lpia/getall\ lpia\ lpia\ lpiacustom/ -i getabis
$ cd ../../
$ cd rules.d/
$ sed -e s/lpia/lpia\ lpiacustom/ -i lpia.mk
$ cd ../
```

Now enter the config directory. This is where all of the specified architecture configuration settings can be modified:

```
$ cd config/lpia/
```

As you can see in this directory, there are two config files. One is the common configuration file `config` for all flavors. Another is `config.lpia`, which is the particular configuration for the `lpia` flavor. As the new flavor is based on the current lpia flavor, you can copy the particular configuration file, as follows:

```
$ cp config.lpia config.lpiacustom
```

Now append the config file for all flavors. This will make it possible to remove the common config file later on, as every flavor will have the configuration.

```
$ cat config >> config.lpia
$ cat config >> config.lpiacustom
```

The next step is to edit the `lpiacustom` configuration file. This can be done manually through the following command:

```
$ vim config.lpiacustom
```

It can also be done by using an ncurses front-end configuration tool for the kernel.

> Ncurses is a toolkit for developing GUI-like applications that run under a terminal. Its header files must be installed prior to proceeding.

First go back to the parent directory:

```
$ cd ../../../
```

Now prepare the kernel and then configure it using the front end:

```
$ debian/rules prepare-lpiacustom
$ make O=debian/build/build-lpiacustom menuconfig
```

At this point, a screen that looks like Figure 8-1 should appear.

Chapter 8: Kernel Fine-Tuning

Figure 8-1

In the new flavor, you will change only some processors that are related options, and all the rest will be kept the same as in the previous "lpia" flavor. So, select the highlighted Processor type and features option that's shown in Figure 8-2.

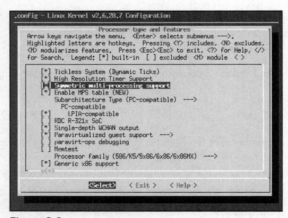

Figure 8-2

Two specific options are going to be customized: Symmetric multi-processing support and Processor family. As the target device is a single processor machine, the first option can be disabled in order to avoid a small overhead generated to deal with multiprocessor machines. In the second case, choosing the correct processor and not the generic one allows for optimization.

The configurations demonstrated here might not be the best for your device. Please refer to your device's technical reference to figure out what options should be selected. When working with an OEM, this should also be specified in an agreement called a "term of work."

So, after those changes are made, the screen will appear as in Figure 8-3.

Chapter 8: Kernel Fine-Tuning

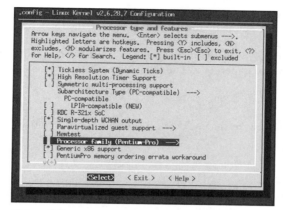

Figure 8-3

Of course, it's also possible to customize any other configuration in this step. Make sure the changes are saved before leaving this front end by selecting Yes, as shown in Figure 8-4.

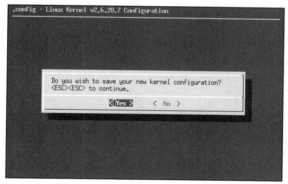

Figure 8-4

Now there is a new customized configuration file. That new file should replace the previous config.lpiacustom:

```
$ cp debian/build/build-lpiacustom/.config \
  debian/config/lpia/config.lpiacustom
```

The next two commands will finish this customization stage. As you already have the particular (and optimized) configuration, and the generic config files were already appended to all the other configuration files in a previous step, you can delete this common file in order to generate a another up-to-date one:

```
$ rm debian/config/lpia/config
```

169

Chapter 8: Kernel Fine-Tuning

Before generating the new common config file, there's an observation to be made depending on how the customization was done. If it's done by the ncurses front end, the following command will run without any interaction from the user. However if it's edited manually, it is possible that some questions will be asked and you should proceed only if you are absolutely sure what is being done. For this reason, we recommend that you don't do it manually. Use the ncurses front end instead:

```
$ debian/scripts/misc/oldconfig lpia
```

The next step is to add the new kernel packages. It is necessary to create the following files:

```
debian/control
debian/control.stub
```

In both files, it's necessary to edit and create three new packages based on the current lpia package. The package entries that are going to be copied are: linux-image-2.6.28-8-lpia, linux-headers-2.6.28-8-lpia, and linux-image-debug-2.6.28-8-lpia. After copying the files, it's necessary to edit them in order to write the correct descriptions. Afterwards, the user should have three new entries.

The first one is:

```
Package: linux-image-2.6.28-8-lpiacustom
Architecture: lpia
Section: base
Priority: optional
Pre-Depends: dpkg (>= 1.10.24)
Provides: linux-image, linux-image-2.6, fuse-module, kvm-api-4,
redhat-cluster-modules
Depends: initramfs-tools (>= 0.36ubuntu6), coreutils | fileutils (>= 4.0),
module-init-tools (>= 3.3-pre11-4ubuntu3), wireless-crda
Conflicts: hotplug (<< 0.0.20040105-1)
Recommends: lilo (>= 19.1) | grub
Suggests: fdutils, linux-doc-2.6.28 | linux-source-2.6.28
Description: Customized Linux kernel image for version 2.6.28
 This package contains a customized Linux kernel image for version 2.6.28.
 .
 Also includes the corresponding System.map file, the modules built by the
 packager, and scripts that try to ensure that the system is not left in an
 unbootable state after an update.
```

The second new entry is:

```
Package: linux-headers-2.6.28-8-lpiacustom
Architecture: lpia
Section: devel
Priority: optional
Depends: coreutils | fileutils (>= 4.0), linux-headers-2.6.28-8, ${shlibs:Depends}
Provides: linux-headers, linux-headers-2.6
Description: Linux kernel headers for version 2.6.28
 This package provides kernel header files for version 2.6.28.
 .
 This is for sites that want the latest kernel headers.  Please read
 /usr/share/doc/linux-headers-2.6.28-8/debian.README.gz for details.
```

Chapter 8: Kernel Fine-Tuning

The third new entry is:

```
Package: linux-image-debug-2.6.28-8-lpiacustom
Architecture: lpia
Section: devel
Priority: optional
Provides: linux-debug
Description: Linux kernel debug image for version 2.6.28
 This package provides a customized kernel debug image for version 2.6.28.
 .
 This is for sites that wish to debug the kernel.
 .
 The kernel image contained in this package is NOT meant to boot from. It
 is uncompressed, and unstripped. This package also includes the
 unstripped modules.
```

Now that all dependencies and steps are completed, the next step is to compile the kernel and generate the new customized kernel packages. Type the command that follows:

```
$ AUTOBUILD=1 NOEXTRAS=1 fakeroot debian/rules binary-lpiacustom
```

At this point, getting a cup of coffee is a nice idea while you wait for the end of the compilation. When this step is finished, it will be possible to find the packages that were created in the parent directory. Go there and analyze all the new files:

```
$ cd ../
$ ls *.deb
```

Now it is completed. All that is left is to install the kernel package files using the following command:

```
$ sudo deb -i <put right files>.deb
```

As mentioned earlier, it is necessary to explain how to get the kernel source code without the need for git version control. As it is not much more simple than when using git, and the downloaded source is almost always an out-of-date Ubuntu kernel version, this is not the better choice. Do this only if you really don't need an up-to-date kernel, or if you just prefer to do it this way.

Before proceeding, deb-src entries must be added to the source.list file.

Next, synchronize the package index file by typing the command:

```
$ sudo apt-get source linux-image-`uname -r`
```

Now it's possible to get the kernel source directly from deb repositories:

```
$ sudo apt-get source linux-image-`uname -r`
```

The directory created is pretty much the same as was created through the git tool, and the steps to customize the kernel are identical as well.

Chapter 8: Kernel Fine-Tuning

Create a Debian Package

If the OEM or user aims to build a non-Ubuntu kernel source or would like to generate a Debian package, this is probably the right method to choose because you want to use the latest official 2.6 kernel tree. Of course, it will also work with any other official and non-official compilable kernel tree.

This method offers two primary ways to get the kernel source code. As in the previous Ubuntu method, it's possible to get it either from git or from the debian repositories. Let's first look at the git repository and then the deb source repositories.

The non-Ubuntu git repository used here is the latest official linux-2.6 kernel tree from kernel.org. However, the source repository could be changed. The same logic applies to the deb repository as well. Only the initial steps will be different — i.e. the kernel configuration, compilation, and installation steps are all kept equal.

The first step in this method is to ensure the user has all dependencies already installed on the device:

```
$ sudo apt-get update
$ sudo apt-get install git-core fakeroot makedumpfile
$ sudo apt-get install libncurses5-dev build-essential
$ sudo apt-get install kernel-package
$ sudo apt-get build-dep linux
```

Now get the linux tree:

```
$ git clone \
  git://git.kernel.org/pub/scm/linux/kernel/git/torvalds/linux-2.6.git \
  linux-2.6.git
```

It's important to mention that this tree is intended for development purposes and may not be stable. So, if anything during the compilation goes wrong, it is possible to change to any previous workable version using the git tool.

This step can last some time depending on the connection speed. After it's completed, a directory called linux-2.6.git will be created. This is the place where all the work will be done. Change to:

```
$ cd linux-2.6.git/
```

Unlike the first method, there is no pre-built workable configuration file. In order to avoid wasting time looking for an initial workable one, just copy the configuration file used by the current kernel. This should decrease the customizing time:

```
$ cp /boot/config-`uname -r` .config
```

Now the kernel source is downloaded and there is an initial configuration file. It is time to start the configuration stage. There are many ways to do this and the ncurses front end will be shown here.

The configuration steps through the ncurses front end will look like Figures 8-1, 8-2, and 8-3, respectively. It can be started by typing the following command:

```
$ make menuconfig
```

Chapter 8: Kernel Fine-Tuning

Before leaving the front end, make sure that everything is saved properly by choosing Yes in the screen that's shown in Figure 8-4.

If the user did not use any front end, before proceeding to the next step, it's interesting to run an extra command to update the configuration file edited:

```
$ make oldconfig
```

Now compile and create the Debian package. We should define two parameters here: `--append-to-version` *name* and `--revision` *number*. The first parameter is a short description of the target's customized kernel. The second one is useful if you want to inform the Debian package manager which is the newer compilation when there are various compiled packages in the future. So, the first compilation will start with 1, and every new package built will have a revision increased by 1.

To make the kernel compile and to ensure that the package is properly built, type the following command:

```
$ fakeroot make-kpkg --initrd --append-to-version -cuustom --revision 1 \
    kernel_image kernel_headers
```

By adding kernel_image and kernel_headers, the kernel image package and its headers are built as well. The `--initrd` is needed to create an initial RAM disk and to make the boot happen in two stages. In the first stage, the initial RAM disk is mounted as a temporary root filesystem, which will help the system's root filesystem to be mounted properly. Thanks to this first stage, the configured kernel could compile as a module (and not as a built in) almost everything in the kernel. These required modules will be loaded before the real filesystem is mounted.

When that compilation step is finished, the kernel packages are ready to be installed. To ensure that this completed correctly, look for the following files in the parent directory:

```
$ cd ../
$ ls linux-image-<kernel's version>-1-custom.deb
$ ls linux-headers-<kernel's version>-1-custom.deb
```

If, for any reason, any problem occurred in the compilation step, go back to any known workable version of the kernel — for example, v2.6.29-rc6 at the time of this writing. Each official version has a tag in the git tree.

Git is explained in further detail in Appendix B but it will be covered here briefly, too. Enter the kernel directory again:

```
$ cd linux-2.6.git/
```

Now type the command that follows to see all tags created in the kernel git tree. Then choose the latest one or any other older one.

173

Chapter 8: Kernel Fine-Tuning

```
$ git tag

(...)
v2.6.28
v2.6.28-rc1
v2.6.28-rc2
v2.6.28-rc3
v2.6.28-rc4
v2.6.28-rc5
v2.6.28-rc6
v2.6.28-rc7
v2.6.28-rc8
v2.6.28-rc9
v2.6.29-rc1
v2.6.29-rc2
v2.6.29-rc3
v2.6.29-rc4
v2.6.29-rc5
v2.6.29-rc6 ← latest
```

As the official version is the newest one, make the git tree go back to it by executing the command shown here:

```
$ git checkout v2.6.29-rc6
```

Before proceeding, it's necessary to update the kernel's configuration file again. It might not be necessary to run any front end now so just type the command:

```
$ make oldconfig
```

For every possible question, just hit Enter and choose the default option. Now, the kernel is ready to be compiled again. Run the following instructions to restart the compilation:

```
$ make-kpkg clean
$ fakeroot make-kpkg --initrd --append-to-version -custom --revision 1 \
  kernel_image kernel_headers
```

If everything is fine this time, it's possible to see the built packages by running the commands demonstrated earlier.

Now, the last step is just to install those packages. Go to the parent directory:

```
$ cd ../
```

The packages can be installed by typing the following:

```
$ sudo dpkg -i linux-image-<kernel's version>-1-custom.deb
$ sudo dpkg -i linux-headers-<kernel's version>-1-custom.deb
```

The alternative way to download the kernel source is via the dpkg tool. Unfortunately, using this method will almost always mean an out-of-date kernel version will be downloaded. The first action here then is to figure out which version is available:

```
$ sudo apt-get install linux-image-2.6
```

The message printed will be similar to this one:

```
Reading package lists... Done
Building dependency tree
Reading state information... Done
Package linux-source-2.6 is a virtual package provided by:
  linux-source-2.6.26 2.6.26-13
  linux-source-2.6.28 2.6.28-1
You should explicitly select one to install.
E: Package linux-source-2.6 has no installation candidate
```

The latest one should be selected:

```
$ sudo apt-get install linux-source-2.6.28
```

When the package is installed, it has to be extracted to any workable directory by the following command:

```
$ cd <user's directory>
$ tar xjf /usr/src/linux-source-2.6.28.tar.bz2
$ cd linux-source-2.6.28
```

The source is now ready.

Updating a Customized Kernel Tree

The kernel will continuously evolve. Newer versions are going to be released after the customized kernel packages created earlier were made. Fortunately, the git control version tool supports updating a modified kernel tree (at least with less effort than reapplying all the changes shown previously).

We will demonstrate two scenarios for updating a kernel tree — the choice depends on whether an Ubuntu Linux kernel tree is being used or not. It is important to have a good understanding of the git tool in order to proceed in this section, particularly if the first method is chosen. Please refer to Appendix B, which covers git usage, if you need clarification for any step.

Updating an Ubuntu Kernel Tree

If you choose the first method, one of the main advantages is that the latest Ubuntu kernel version can be used. If new changes are released in that git tree, it's possible to update the downloaded kernel tree and apply it without having to redo the modifications.

Chapter 8: Kernel Fine-Tuning

The first step when doing this is to verify the modified files and the current branch in the git tree. Before doing anything go to the kernel tree directory. Then start by typing the following command to get the current branch:

```
$ git branch
```

The result should be like this:

```
$ * master
```

This means the "master" branch is selected. Now, get the list of all modified files:

```
$ git status
# On branch master
# Changed but not updated:
#   (use "git add <file>..." to update what will be committed)
#
#       modified:   debian/config/lpia/config
#
#       modified:   debian/config/lpia/config.lpia
#
#       modified:   debian/control
#
#       modified:   debian/control.stub
#
#       modified:   debian/rules.d/lpia.mk
#
#       modified:   debian/scripts/misc/getabis
#
# Untracked files:
#   (use "git add <file>..." to include in what will be committed)
#
#       build/
#
#       debian/config/lpia/config.lpiacustom
```

There are two lists in the result: "Changed but not updated" and "Untracked files." The first one refers to all files that already exist but are modified. The second one refers to the new files (either created by the user or not). It's possible more untracked files could be found than those shown in the preceding code. That should not be a problem. However if the *exact* steps were followed from Creating An Ubuntu Package, the first list should be as shown. If not, redo this section.

To save the new customized configuration, a new branch needs to be created in the user's git tree — and then the "master" branch should be reset to its locally unmodified state. Let's create the new branch called "custom" and check it out:

Chapter 8: Kernel Fine-Tuning

```
$ git checkout -b custom
M       debian/config/lpia/config
M       debian/config/lpia/config.lpia
M       debian/control
M       debian/control.stub
M       debian/rules.d/lpia.mk
M       debian/scripts/misc/getabis
Switched to a new branch "custom"
```

The new branch is now created. The modified files need to be added prior to storing their modifications:

```
$ git add debian/config/lpia/config debian/config/lpia/config.lpia
$ git add debian/control debian/control.stub debian/rules.d/lpia.mk
$ git add debian/scripts/misc/getabis
```

Then, the untracked file called config.lpiacustom, which was created previously, needs to be added using the following command:

```
$ git add debian/config/lpia/config.lpiacustom
```

All modifications are now marked to be saved. It's possible to check this by typing the following command:

```
$ git status
```

And the result should be:

```
# On branch custom
# Changes to be committed:
#   (use "git reset HEAD <file>..." to unstage)
#
#       modified:   debian/config/lpia/config
#       modified:   debian/config/lpia/config.lpia
#       new file:   debian/config/lpia/config.lpiacustom
#       modified:   debian/control
#       modified:   debian/control.stub
#       modified:   debian/rules.d/lpia.mk
```

Chapter 8: Kernel Fine-Tuning

```
#       modified:   debian/scripts/misc/getabis

# Untracked files:

#   (use "git add <file>..." to include in what will be committed)

#       build
```

Unlike the previous "git status," this time there's a new list called "Changes to be committed." Every modification that's included in this list will be committed using the following command:

```
$ git commit
```

A text editor appears. The command will complete after the file is saved and the editor is exited. If the editor is exited without saving, the commit command is aborted. Before saving the file, the user needs to write some short description for the commit, as shown in the following code:

```
lpia: custom: Customized configuration for lpia

This patch adds the custom flavour
 in lpia arch

Signed-off by: User Name <user@email.com>
# Please enter the commit message for your changes. Lines starting
# with '#' will be ignored, and an empty message aborts the commit.
# Committer: root <root@esdhcp03540.(none)>
# On branch custom
# Changes to be committed:
#   (use "git reset HEAD <file>..." to unstage)
#       modified:   debian/config/lpia/config
#       modified:   debian/config/lpia/config.lpia
#       new file:   debian/config/lpia/config.lpiacustom
#       modified:   debian/control
#       modified:   debian/control.stub
#       modified:   debian/rules.d/lpia.mk
#       modified:   debian/scripts/misc/getabis
# Untracked files:
#   (use "git add <file>..." to include in what will be committed)
#       build/
```

This message is similar to an e-mail but with some additional rules. The first one is that all lines starting with # will be ignored. The first line is the subject. Its description should be short and direct. Then, after an empty line, is the patch's description as the e-mail's body. It needs to be longer and more complete and can have more than one line. After one more empty line is the patch's sign-off. The person who signs the "patch-off" guarantees that it really works. After saving the file and exiting the editor, all modifications are applied to the local working branch. By typing the next command, it's possible to confirm that the patch is really saved:

```
$ git log
```

Chapter 8: Kernel Fine-Tuning

The resulting screen looks as follows:

```
commit 6d6ef0fca469453cc64a6bcfca31e15a1c8a002f

Author: root <root@esdhcp03540.(none)>

Date:   Sun Mar 22 14:57:59 2009 +0000
    lpia: custom: Customized configuration
    This patch adds the custom flavour in lpia arch
    Signed-off by: User Name <user@email.com>
commit 75d2d4cd02f92e7c7d9fce28bef556c79bdbed92
Author: Amit Kucheria <amit.kucheria@ubuntu.com>
Date:   Thu Feb 26 12:00:37 2009 +0200
    UBUNTU: Updating configs (arm:ixp4xx)
    Beeper is compiled in to beep helpfully during installs
    Signed-off-by: Amit Kucheria <amit.kucheria@ubuntu.com>
commit 68a81f84a1361bc016209dfadd1f0ad1371dedf3
Author: Tim Gardner <tim.gardner@canonical.com>
Date:   Wed Feb 25 14:25:05 2009 -0700
    UBUNTU: Start new release
    Ignore: yes

    Signed-off-by: Tim Gardner <tim.gardner@canonical.com>
(More commits...)
```

There's no need for it to be exactly the same as just shown. Probably the last two entries will not be the same and the first one could have some differences as well. The main issue here is that the first description must be the same as what was entered previously. If it is, then it is done. The new branch's status can be obtained from git as a confirmation that there's no pending modification:

```
$ git status
```

```
root@esdhcp03540:~/ubuntu-jaunty.git# git status



# On branch custom

# Untracked files:

#   (use "git add <file>..." to include in what will be committed)

#       build/
```

All files added prior to the commit are not listed anymore. This means that their modification is now part of the local branch. So, as the next step, let's check out the "master" branch again to be able to update it and get the latest Ubuntu kernel version.

```
$ git checkout master
```

Chapter 8: Kernel Fine-Tuning

Before proceeding with the kernel update, it's necessary to make sure there are no pending modifications in the working branch. You can once again use the `git status` command:

```
$ git status
```

If there is a "Changed but not updated" list, then the working branch is not clean. It's possible to reset the working branch and lose all modification with the following command:

```
$ git reset -hard HEAD
```

Now the working branch does not have any pending modifications. The next step is to update the "master" branch:

```
$ git pull
```

A lot of messages will be printed out now. It is only necessary to be worried if an error message is shown. If it is, the best action is to try and reset, and pull again. This branch is a copy of the official Ubuntu branch, so the patch is only applied against the "custom" one. To achieve the target of this section and get the "custom" branch up-to-date, let's check out the "custom" branch again:

```
$ git checkout custom
```

The next git command will update the "custom" branch and then try to reapply the user's patch on top of that:

```
$ git rebase master
```

The result now is unpredictable and there are many ways to solve any possible conflicts or problems. If everything is fine, just proceed to the next step; if a conflict occurs, a solid knowledge of git is fundamental. Please refer to Appendix B. If a conflict occurs that you can't solve, the best solution is to reset the file to the "master" branch and then redo the changes afterwards.

To figure out if there's a conflict, see if a message like this one (containing the error entries) is printed out after the `rebase` operation:

```
First, rewinding head to replay your work on top of it...
Applying: 1
error: patch failed: debian/control:285
error: debian/control: patch does not apply
error: patch failed: debian/control.stub:285
error: debian/control.stub: patch does not apply
error: patch failed: debian/scripts/misc/getabis:77
error: debian/scripts/misc/getabis: patch does not apply
```

Then, for each file with an error, just type the following:

```
$ git checkout master <error file>
$ git add <error file>
```

In the case of conflicts, the following command will continue the `rebase` operation:

```
$ git rebase --continue
```

If a file were reset to the "master" branch, then is should be modified again. After it's done, or if no conflict occurred, then the kernel is now up-to-date.

Update a Non-Ubuntu Kernel Tree

The second update method is really simple to execute. It is best used if the second method was utilized to create a kernel package. There are two main steps: updating the kernel tree and updating the kernel configuration.

First update the kernel tree by typing the following command:

```
$ git pull
```

After the command is done, the next step is to update the kernel configuration in the existing .config file in the kernel's directory. As demonstrated, it's possible to configure the kernel either using an ncurses front end or using the very simple text interface. The ncurses front end is selected with the following command:

```
$ make menuconfig
```

It's possible now to configure the kernel. Another alternative configuring method is as follows:

```
$ make oldconfig
```

This one is a bit more complicated as a query pops up for each new configuration added to the kernel. The front end will automatically choose the default value for every new item, but the same can be done in the simple text interface by simply pressing Enter for all the questions.

Dynamic Kernel Module Support

Refer to the Dynamic Kernel Module Support (DKMS) official page for more information, which you'll find at http://linux.dell.com/dkms/.

As shown previously, it is possible to tune a Linux kernel to make the configuration closer to the actual device hardware being used, rather than sticking with a more generic solution provided by the distribution. According to the methods shown previously, the entire kernel needs to be compiled and installed. This means that the tuned kernel could be quite different from the generic one and could behave quite differently, too. Of course, it might not be a problem if the source comes from a stable release.

But what if the OEM just needs to update a driver version or just use a drive that is not present in the current kernel? Is it necessary to change the whole kernel to a new (or old) version that contains the required driver? The answer is no, thanks to the Linux kernel's ability to add new functionalities at runtime by loading a kernel module. It is possible to compile and load a new driver without having to touch the already configured and compiled kernel image.

There are many reasons for the OEM not to want to change the current kernel when trying to use a new or up-to-date driver version:

Chapter 8: Kernel Fine-Tuning

- It's much easier and quicker to compile and install the driver only.
- The OEM may need to stick with the current kernel for compatibility reasons.
- In some situations, it might be forbidden to use a non-tested kernel for reliability reasons. This means that a new kernel version would result in more enforced validation and increased costs before being pushed to an end user.
- The required driver (or driver version) is not currently present in the kernel tree.

There are a plenty of different ways to deliver a driver that resides outside the kernel tree or a new version that is currently either not applied or not available in the current kernel yet. If it's not well standardized, it can mean a lot of work for driver developers to make each delivered driver version compatible with the many kernel versions and/or Linux distributions that exist and also a lot of work for the final users to get, build, and install the driver.

In order to help developers and users, a framework that's called *Dynamic Kernel Modules Support* was created. It provides a very easy mechanism to deliver and install new (or up-to-date) drivers by using few and simple commands and configurations. The following section demonstrates how to use this framework and how users can get the most out of it.

Inside the DKMS Framework

The DKMS has some commands to export its usability to the *userland*. But before we demonstrate them, it's important to show how this framework works internally.

"Userland" refers to any code that's located outside the kernel.

DKMS has its own modules tree. Every module needs to be added to that tree prior to being built and properly installed. The four valid states a module can be in are as follows:

- Out of DKMS tree
- Added to DKMS tree
- Built
- Installed

The desired module needs to reside in the /usr/src path inside the <module>-<version>/ directory. That is the place where the framework will look for the drivers. The module must have a dkms.conf file properly configured. This configuration file should be provided by whoever is delivering the driver with DKMS framework support.

In order to better understand DKMS usage, imagine the following scenario. There is a module package called openafs, which uses DKMS and comes natively with Ubuntu. This is a distributed, cross-platform, open source filesystem for sharing of files between remote computers. Currently, the version provided

Chapter 8: Kernel Fine-Tuning

by Ubuntu is 1.4.9, but the latest stable version is 1.4.10. We'll use DKMS to demonstrate how to install the up-to-date version in a pre-installed tuned kernel, while at the same time keeping the openafs Ubuntu version installed in the generic kernel provided by the distribution. Consequently, both versions will be installed at the same time and both managed by the DKMS framework.

Basic DKMS Commands

There will be two kernels installed: linux-image-2.6.28-11-generic (Ubuntu) and linux-image-6-29-custom (Tuned).

It's important to note that the headers must also be installed on the kernel images.

The kernel from Ubuntu will be the active kernel. The first step is to install the openafs-1.4.9 already provided by Ubuntu. Notice that everything should be done as root.

```
# apt-get install openafs-modules-dkms
```

As you can see, the openafs module is automatically compiled and installed. This happens because the `dkms` commands are already encapsulated in the .deb installation scripts. Now that the 1.4.9 version has been installed, it is necessary to download the newer version (1.4.10) so that this can be installed alongside the 1.4.9 version. The more recent version is already present in the Debian repositories, so the next step is to get it:

```
# wget -c http://ftp.debian.org/debian/pool/main/o/openafs/openafs-modules-dkms_1.4.10+dfsg1-2_i386.deb
```

It is not advisable to install the driver as a regular deb file. If this were to happen, it would overwrite the Ubuntu package and consequently overwrite the older version. The goal is to keep the Ubuntu version with the Ubuntu kernel and install the newest version in the tuned kernel.

Inside the deb file is a directory called openafs-1.4.10 which is what is needed. Extract it from the deb file and copy it to the correct path in order to use it with DKMS:

```
# ar vx openafs-modules-dkms_1.4.10+dfsg1-2_i386.deb
# tar xzvf openafs-modules-dkms_1.4.10+dfsg1-2_i386.deb
# cp -fr usr/src/openafs-1.4.10 /usr/src
```

After running these commands, the newest driver is copied to the /usr/src path. At this point the driver is out of the DKMS tree. Now it is possible to start to use the `dkms` commands. The first task is to verify the current status of all the drivers in the tree. At least the openafs v1.4.9 (which is already installed) will be shown:

```
# dkms status
openafs, 1.4.9, 2.6.28-11-generic, i686: installed
```

If nothing is shown after the `status` command, then no openafs module is installed. Something might be wrong with the openafs-modules-dkms package installation. That step must be redone.

Now it is possible to add the up-to-date driver to the tree and change its state to Added to the DKMS tree:

Chapter 8: Kernel Fine-Tuning

```
# dkms add -m openafs -v 1.4.10 -k 2.6.29-custom
```

Like `status`, `add` is the command that tells what you're doing.

Three options can be used in the command:

```
# dkms <command> OPTIONS:
```

- `-m`: Refers to module name
- `-v`: Refers to the module's version
- `-k`: Refers to the kernel version

Now it is possible to repeat the previous step to see what is different:

```
# dkms status
openafs, 1.4.10: added
openafs, 1.4.9, 2.6.28-11-generic, i686: installed
```

As the driver is now added to the tree, it is possible to build the driver with the command:

```
# dkms build -m openafs -v 1.4.10 -k 2.6.29-custom
```

And the `dkms` status is now:

```
# dkms status
openafs, 1.4.10, 2.6.29-custom, i686: built
openafs, 1.4.9, 2.6.28-11-generic, i686: installed
```

With the driver already built, the last step is to install it:

```
# dkms install -m openafs -v 1.4.10 -k 2.6.29-custom
```

Then, the `dkms` status prints it in the Installed state:

```
# dkms status
openafs, 1.4.10, 2.6.29-custom, i686: installed
openafs, 1.4.9, 2.6.28-11-generic, i686: installed
```

This means the two driver versions are both installed; each one is installed in a different kernel following the initial plan.

Because it's possible to build/install modules by using simple DKMS commands, it's also possible to remove/uninstall them. The uninstall command changes an installed driver to the Built state. The driver is not present in the system anymore but it can be installed again without having to rebuild it:

Chapter 8: Kernel Fine-Tuning

```
# dkms uninstall -m openafs -v 1.4.10 -k 2.6.29-custom
# dkms status
openafs, 1.4.10, 2.6.29-custom, i686: build
openafs, 1.4.9, 2.6.28-11-generic, i686: installed
```

The `dkms remove` command results in changing the driver state to Out of the DKMS tree. After that, it's necessary to add/build/install the driver once more:

```
# dkms remove -m openafs -v 1.4.10 -k 2.6.29-custom
# dkms status
openafs, 1.4.9, 2.6.28-11-generic, i686: installed
```

It is very simple to use the DKMS framework as it's shown here. However, this is not all that the framework can do. In order to demonstrate this, let's extend the example scenario. What if the OEM has its custom kernel installed in many devices and it's necessary to install the up-to-date driver in all of them? This would be a considerable job doing each device separately. However, DKMS can create a tarball file from the built driver and load it again from another device that has the same installed kernel. The package goes directly to the Build state and can be installed without the compilation step. It means the extra devices do not need to have the linux-header package (this is necessary *only* for the compilation step).

The driver should be at least in the Build state. So, for the following tests, the openafs-1.4.10 driver must be at least compiled with the `dkms build` command. The OEM can redo the steps if it's necessary. Then, the tarball file can be created by typing the following:

```
# dkms mktarball -m openafs -v 1.4.10 -k 2.6.29-custom
Marking modules for 2.6.29-custom (i686) for archiving...
Marking /var/lib/dkms/openafs/1.4.10/source for archiving
Tarball location:
/var/lib/dkms/openafs/1.4.10/tarball/openafs-1.4.10-kernel2.6.29-custom-i686.dkms.
tar.gz
# cp /var/lib/dkms/openafs/1.4.10/tarball/openafs-1.4.10-kernel2.6.29-custom-i686.
dkms.tar.gz .
```

The file created can be copied and can be used on any other device that has the custom kernel installed. If the device does not have the custom kernel installed but the OEM wants to test the functionality, it can be simulated by removing the driver with the `dkms remove` command, which deletes the /usr/src/openafs-1.4.10 directory. Consequently, there won't be any trace of the module in the current machine. The tarball loading action can be done by typing the following:

```
# dkms ldtarball --archive openafs-1.4.10-kernel2.6.29-custom-i686.dkms.tar.gz
# dkms status
openafs, 1.4.10, 2.6.29-custom, i686: built
openafs, 1.4.9, 2.6.28-11-generic, i686: installed
```

As you can see by the `dkms status`, the driver is already loaded in the Build state. It's necessary just to install it again to get it usable in the current custom kernel.

Chapter 8: Kernel Fine-Tuning

Summary

In a generic kernel package, it's very hard to provide everything that an OEM needs to provide if it is targeting a large, worldwide user base. Even if this were possible, the optimizations come at a cost. This chapter has shown methods for making a kernel closer to a device by getting a generic kernel and optimizing it.

Of course, it is not possible to say exactly what should be done for each device as each device needs a different optimization. Instead, this chapter demonstrated how optimization can be achieved through simple steps. By using a well-tuned kernel, an OEM and, more importantly, an end user can get much more use and pleasure out of a device.

Testing and Usability

Ubuntu Mobile has been developed in a highly decentralized manner. However, in order for it to be attractive to both the ISV and OEM communities, it must provide better or equivalent levels of reliability than that provided by proprietary platforms.

> *Independent software vendor (ISV) is a business term for companies specializing in making or selling software that is designed for mass markets or for niche markets.*

This chapter first focuses on the Mago that the Ubuntu QA team is using to test automate the desktop and which can be used to test devices using the accessibility framework. Next, it shows how to create and test an example application and how to verify its compatibility with the Linux Standard Base (LSB) Testing Toolkit.

The chapter discusses other testing tools — specifically those tools for package testing — and then it shows some tools that are useful for general performance and usability. Finally, the chapter concludes with a discussion about possible strategies for testing. It also shows how to submit a bug report correctly to the Ubuntu developers if errors are discovered.

Why Test?

Testing can never completely establish the correctness of software. Instead, it provides a comparison that measures the state and behavior of the software or hardware against some known or expected benchmarks. These may include specifications, comparable products, past versions of the same product, user or customer expectations, or relevant standards. The need for high quality testing is obvious — it has been estimated that software bugs cost the US economy alone more than $59.5 billion in 2002 (`http://www.nist.gov/public_affairs/releases/n02-10.htm`). It has also been estimated that more than a third of this cost could have been avoided if better software testing was performed.

Chapter 9: Testing and Usability

Ubuntu Desktop QA

Currently, testing the functionality of the GNOME desktop on Ubuntu means using the accessibility libraries to trigger the user interface widgets of the application under test. A prerequisite to this happening is that the Assistive Technology Service Provider Interface (AT-SPI) be enabled.

> *AT-SPI technologies are currently migrating to D-Bus. See Chapter 4 and Appendix E for more on D-Bus.*

To do this on Ubuntu Netbook Remix, go to Preferences ⇨ Assistive Technologies and select the Enable assistive technologies checkbox. This can be seen in Figure 9-1.

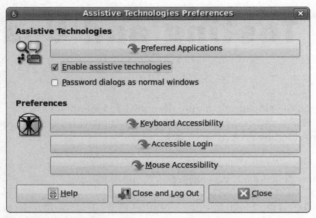

Figure 9-1

> *The interface is enabled at login using gdm, so restart the GNOME session (log out and log in again). Also, if Compiz is enabled for the desktop then accessibility will not work. In this case, choose the plain gnome session in gdm.*

Next, install the Python packages python-ldtp, python-distutils-extra, and python-setuptools (as well as Bazaar if it is not already installed):

```
$ sudo apt-get install python-ldtp, python-distutils-extra python-setuptools bzr
```

Mago — A Desktop Testing Initiative

To use Mago, check out this branch:

```
$ bzr branch lp:mago
```

Move into the newly created folder and run one of the tests:

```
$ cd mago
$ PYTHONPATH=. ./bin/mago -a notify-osd
```

This runs the notify-osd test, which verifies the functionality of the new messaging daemon (for more on this new messaging system in Ubuntu, see Chapter 4). The results of the test are stored in the .mago folder in the home directory, and the HTML file can be displayed in a browser.

Building an Application for Testing

To set up a hypothetical test scenario, an example application will be built and then tested. This application is an embedded web browser, built using the Mozilla libraries. This application will be tested for both functionality and for its compliance with established Linux standards.

Such an embedded browser could be useful for kiosk applications where perhaps the full functionality of a browser is not required.

For the purposes of this chapter, the focus is on keeping the browser simple; some of the GTK widgets in the actual application, such as Refresh and Stop, are not hooked up to anything in terms of code. However, the location bar is functional so that URLs can be entered and URI resources can be accessed.

A URI (Uniform Resource Identifier) consists of a string of characters used to identify or name a resource on the Internet.

Getting Started

In a terminal window, make sure that the required libraries can be imported:

```
$ python
Python 2.5.2 (r252:60911, May  7 2008, 15:19:09)
[GCC 4.2.3 (Ubuntu 4.2.3-2ubuntu7)] on linux2
Type "help", "copyright", "credits" or "license" for more information.
>>> import gtk
>>> import gnome.ui
>>> import gtk.glade
>>> import gtkmozembed
location: /usr/lib/xulrunner-1.9.0.1/libxpcom.so
before 3
>>>
```

gtkmozembed is available in the python-gnome2-extras package.

Chapter 9: Testing and Usability

Application Creation

Start Glade and create an interface with a window, a vertical box, and a toolbar. For each item, set the accessibility name and description. This allows Mago to test the user interface using the AT-SPI accessibility layer bindings.

The Glade widget tree is shown as well as the accessibility properties for the main window in Figure 9-2.

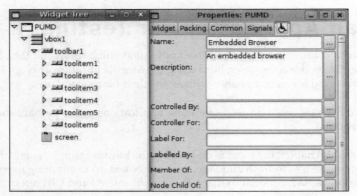

Figure 9-2

The URL entry box needs to be made active. To do this, click on the text entry to make it active in the properties window. Then select the Signals tab and select Activate. A text entry emits the Activate signal when the Enter key is pressed. This will make a default handler of `on_entry1_activate`, which is fine.

The following code snippet displays the beginning of the Glade file. It shows some of the main structures such as `GtkWindow` and `GtkVBox` enabled for accessibility and, therefore, for automated testing:

```
<?xml version="1.0" standalone="no"?> <!--*- mode: xml -*-->
<!DOCTYPE glade-interface SYSTEM "http://glade.gnome.org/glade-2.0.dtd">
<glade-interface>
<widget class="GtkWindow" id="PUMD">
  <property name="visible">True</property>
  <property name="title" translatable="yes">Embedded Browser</property>
  <property name="type">GTK_WINDOW_TOPLEVEL</property>
  <property name="window_position">GTK_WIN_POS_NONE</property>
  <property name="modal">False</property>
  <property name="resizable">True</property>
  <property name="destroy_with_parent">False</property>
  <property name="icon">pixmaps/hello-world-48x48.png</property>
  <property name="decorated">True</property>
```

Chapter 9: Testing and Usability

```xml
      <property name="skip_taskbar_hint">False</property>
      <property name="skip_pager_hint">False</property>
      <property name="type_hint">GDK_WINDOW_TYPE_HINT_NORMAL</property>
      <property name="gravity">GDK_GRAVITY_NORTH_WEST</property>
      <property name="focus_on_map">True</property>
      <property name="urgency_hint">False</property>
      <accessibility>
        <atkproperty name="AtkObject::accessible_description" translatable="yes">
An embedded browser</atkproperty>
      </accessibility>
      <child>
        <widget class="GtkVBox" id="vbox1">
          <property name="visible">True</property>
          <property name="homogeneous">False</property>
          <property name="spacing">0</property>
          <accessibility>
            <atkproperty name="AtkObject::accessible_name" translatable="yes">vbox1
            </atkproperty>
            <atkproperty name="AtkObject::accessible_description" translatable="yes">
            The main box</atkproperty>
          </accessibility>
          <child>
```

Note the accessibility node in the preceding code, which is represented by the following:

```xml
            <accessibility>
               <atkproperty name="AtkObject::accessible_name" translatable="yes">
                  vbox1</atkproperty>
               <atkproperty name="AtkObject::accessible_description" translatable="yes">
                  The main box</atkproperty>
            </accessibility>
```

The browser itself is a wrapper that will initialize the GNOME application and load all of the widgets through a call to `self.widgets = gtk.glade.XML("pumd.glade")`. The following is the code for the browser:

```python
#!/usr/bin/env python
import gtk
import gnome.ui
import gtk.glade
import gtkmozembed
APPNAME="PUMD"
APPVERSION="0.1"

class BrowserWrapper:

    def __init__(self):
        # register as a Gnome application.
```

Chapter 9: Testing and Usability

```
            gnome.init(APPNAME, APPVERSION)
            # load up the glade file
            self.widgets = gtk.glade.XML("pumd.glade")
            # create a widget to hold the browser
            self.mozillaWidget = gtkmozembed.MozEmbed()
            # add the widget to the program
            self.widgets.get_widget("screen").add(self.mozillaWidget)
            self.mozillaWidget.set_size_request(800,600)
            # show the widget
            self.mozillaWidget.show()
            self.mozillaWidget.load_url("wiki.ubuntu.com/MobileAndEmbedded")

            signalDic = {"on_entry1_activate" : self.on_entry1_activate}
            self.widgets.signal_autoconnect(signalDic)

      def on_entry1_activate(self, widget):
            self.mozillaWidget.load_url(widget.get_text())
# if __name__ magic if this is the first module called
if __name__ == "__main__":
    widgets = BrowserWrapper()
    gtk.main()
```

This example program will not exit correctly. You can kill the window easily enough, but you still need to use Ctrl+C to kill the program.

Put this file in a folder and then create a subfolder called pixmaps with the icon for the application; the icon can be changed in the Glade file with the following line:

```
<property name="icon">pixmaps/hello.png</property>
```

Run the browser by making the Python file executable and calling it with the interpreter:

```
$ chmod +x pumd.py
```

```
$ python pumd.py
```

It is not really necessary to call the Python interpreter here. The preceding command could just as easily be replaced by ./pumd.py.

Now that the browser is working, it is possible to test it. To do this, the browser will be tested for functionality with Mago and for compatibility with the Linux Standard Base. The reason that compatibility is important is that it links development more closely to certification with the result being reduced development costs and a tighter integration between applications and the LSB standard.

Chapter 9: Testing and Usability

Testing with Mago

Make sure that Mago has been branched from Launchpad (see the preceding text)

Adding the Browser Test to Mago

From the root folder of the Launchpad Mago checkout, edit the file mago/test_suite/ubuntu.py and add the following:

```
class PumdTestSuite(SingleApplicationTestSuite):

    APPLICATION_FACTORY = PUMD
```

This registers the new PUMD class as a new suite. Next, add the code for this PUMD class, which will initialize the application and provide some utility functions. The code is added to mago/application/ubuntu.py. First import some libraries:

```
from time import  sleep

import tempfile
```

and then add the class. Notice that this is a child of the Application class and the full path to the browser executable.

```
class PUMD(Application):

    MNU_ITEM      = "mnuEmbeddedBrowser"
    WINDOW        = "frmEmbeddedBrowser"
    LAUNCHER      = "/home/ian/Dev/Book/testing/mago/pumd/browser/pumd.py"

    def __init__(self):

        """
        Embedded Browser class init method
        """
        self.screenshots = []
        Application.__init__(self)

    def open(self):

        """
        This opens the Embedded Browser application and raises an error if the
        application didn't start properly.
        """
        try:
            Application.open(self)
        except ldtp.LdtpExecutionError:
```

Chapter 9: Testing and Usability

```
                raise ldtp.LdtpExecutionError, "The application did not start
correctly."

    def grab_image_and_wait(self):
        sleep(1)
        screenshot = \
                ldtputils.imagecapture(outFile=tempfile.mktemp('.png', 'pumd_'))
        self.screenshots.append(screenshot)
        return (screenshot)
```

With the test suite registered and the class written, it is now possible to actually use it to write a test for the browser. This is done by creating a folder in the root of the Mago checkout (named after the application to be tested) and by adding a Python and an XML file to this folder. The Python file looks like this:

```
from mago.test_suite.ubuntu import PumdTestSuite
from time import sleep

class PUMD(PumdTestSuite):
        def testOpenMenu(self, menuitem=None, windowname=None, closetype=None,
closename=None, oracle=None):
                self.application.set_name(windowname)
                self.application.open()
                sleep(20)
                screeny = self.application.grab_image_and_wait()
                try:
                        screeny
                except Exception, e:
                        print "No screenshot returned"
                        raise e
if __name__ == "__main__":
    pumd_test = PUMD()
    pumd_test.run()
```

It is important to notice the import of the `PumdTestSuite`; the `run()` method, which calls the test runner in the suite; and the `sleep()` method, which gives the page time to load. The call to `grab_image_and_wait()` drops a screenshot — shown in Figure 9-3 — into the temp folder.

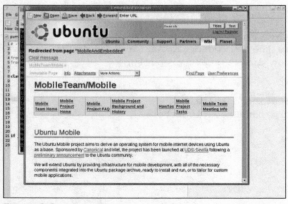

Figure 9-3

Chapter 9: Testing and Usability

The XML file looks as follows:

```xml
<?xml version="1.0"?>
<suite name="pumd">
  <class>pumd.PUMD</class>
  <description>
    Test which verifys that the browser starts correctly.
  </description>
  <case name="Open Embedded Browser">
    <method>testOpenMenu</method>
    <description>It opens the Embedded Browser</description>
    <args>
      <menuitem>mnuEmbeddedBrowser</menuitem>
      <windowname>frm*EmbeddedBrowser</windowname>
    </args>
  </case>
</suite>
```

Also notice that the `<class>` tag contains the module name of the .py file and the `SuiteClass` class. Run the test by using the following:

```
PYTHONPATH=. ./bin/mago -a pumd
```

The results — shown in Figure 9-4 — will be placed in the ~/.mago/pumd folder.

Figure 9-4

This shows that the browser open test passed, and it shows the time it took to run the test. If the test fails, then a screenshot is taken and placed in the ~/.mago/pumd/screenshots folder and the error is logged into the ~/.mago/pumd/pumd.log file — both of which help when debugging.

Chapter 9: Testing and Usability

Linux Standards Base and Certification

The LSB Application Testkit (ATK) contains tools for analyzing dependencies (libraries and interfaces that are required externally) of application packages. In particular, ATK helps developers test their applications for LSB compliance, and enables easy steps for LSB certification.

Installing the LSB Application Testkit

First, download the toolkit from http://www.linuxfoundation.org/en/Downloads. Next, run the install.sh script in the unpacked directory. The script asks a couple of questions, and installs the packages using dpkg:

```
This system appears to be a variant of Debian GNU/Linux, such as
Debian itself, Ubuntu, MEPIS, Xandros, Linspire, etc.
Is this correct? yes
In order to install these packages, you need administrator
privileges.  You are currently running this script as an unprivileged
user.
You have sudo available.  Should I use it? yes
Using the command "sudo /bin/sh -c" to gain root access.  Please type the
appropriate password if prompted.
Installing packages...
```

Running the LSB Application Testkit

The toolkit installs into /opt/lsb/app-testkit/manager; to start the web interface, run /opt/lsb/app-testkit/manager/bin/lsb-atk-start.pl 8080 and then browse to http://127.0.0.1:8080/.

This will display the test suite that's shown in Figure 9-5.

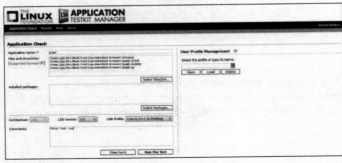

Figure 9-5

Chapter 9: Testing and Usability

Open the Application Check page, and specify the name of the application and the file to test pumd.py.

Next, select the LSB version and profile (the `Architecture` field cannot be changed currently), and press the Run the Test button.

The test will be executed, displaying the Test Report page. It contains a lot of information about the test run.

From this page, you can start the LSB certification process for the application you have just tested.

In the test, the toolkit recommended changing the embedded browser

```
#!/usr/bin/env python
```

to

```
#!/usr/bin/python
```

This recommendation was made in order to increase the portability of the application across different platforms. It can be seen then that the LSB aims to maintain compliance across architectures and, as such, it is an emerging standard in the open source ecosystem. Having an application certified by the LSB is highly recommended and indeed perhaps becoming mandatory for most ISV's.

Other Testing Tools

In addition to Mago and LSB, the open source community has developed a range of other testing applications. These include Phoronix Test Suite, which is a testing and benchmarking platform; pbuilder, which is used for package testing; and a range of other tools (such as ps and top), which come standard with the Ubuntu operating system.

Using pbuilder is explained in Chapter 5.

Phoronix Test Suite

The Phoronix Test Suite is designed to carry out both qualitative and quantitative benchmark testing of a Linux operating system. It consists of a processing core (*pts-core*) and benchmarks that are XML-based profiles with related resource scripts. You can find the application in the universe repositories. You can install it using apt:

```
$ sudo apt-get install phoronix-test-suite
```

Chapter 9: Testing and Usability

To see all available tests, run the following:

```
$ phoronix-test-suite list-tests
```

To run a test, use the following:

```
$ phoronix-test-suite benchmark <test>
```

To run the netbook test, for example, use

```
$ phoronix-test-suite benchmark netbook
```

The application will automatically download all the necessary test files, and will prompt for test options and whether the test results should be saved.

The suite also supports monitoring system sensors while the test is running. At the end of the test, it will provide the low and high thresholds for each sensor as well as the average. The sensor detection and monitoring is done through lm-sensors and the ACPI interface:

```
$ sudo apt-get install lm-sensors
```

lm-sensors can be enabled by running the following:

```
$ sudo sensors-detect
```

For a practical example of the use of sensors, see Chapter 3.

The suite was used to benchmark the released i386 version of Ubuntu Netbook Remix for Ubuntu Jaunty against an LPIA version, which was made for this book by installing the Netbook Remix desktop over a MID base.

The benchmark that was run on both was the netbook suite, and the command that was used in both tests was:

```
$ MONITOR=all   phoronix-test-suite benchmark netbook
```

The hardware under test was an Acer Aspire with the following specifications:

```
Processor: Intel Atom CPU N270 @ 1.60GHz (Total Cores: 2), Motherboard:
Acer AOA110, Chipset: Intel Mobile 945GME Express Hub + ICH7-M,
System Memory: 485MB, Disk: 8GB SSDPAMM0008G1,
Graphics: Intel Mobile 945GME Express IGP (rev 03)
```

The software installed was:

```
OS: Ubuntu 9.04, Desktop: GNOME 2.26.1, Display Server: X.Org Server 1.6.0,
Display Driver: intel 2.6.3
```

Chapter 9: Testing and Usability

The difference between the two systems under test can be seen in the kernel:

```
Kernel: 2.6.28-11-generic (i686)
```

which is the default on the released version of the Ubuntu Netbook Remix and

```
Kernel: 2.6.28-11-lpia (i686)
```

which is the default on the LPIA version. The test results are presented in the following table.

Test Detail	UNR Default	UNR LPIA	% Difference
LAME MP3 Encoding (encode an mp3 file)	163.21	175.22	UNR Default better by 7.35%
OGG Encoding (encode an ogg file)	102.05	105.79	UNR Default better by 3.66%
Timed ImageMagick Compilation (Time taken to compile Image Magick)	906.16	912.02	UNR Default better by 0.64%
512MB Write Performance	5.40	5.31	LPIA better by 1.69%
512MB Read Performance	33.00	33.08	UNR Default better by 0.24%
1GB Write Performance	5.58	5.48	LPIA better by 1.82%
OpenSSL RSA 4096-bit Performance (Encryption test on SSL)	4.50	4.30	LPIA better by 4.65%
GnuPG 2GB File Encryption (Encrypt a 2GB file with GnuPG)	206.17	158.72	LPIA better by 29.89%
RAM speed (Writes Integers to RAM)	1834.82	1708.05	UNR Default better by 7.42%

The most obvious difference is that file encryption on LPIA is nearly one-third faster than on the default i386. Of the nine tests that were run, the default Ubuntu Netbook Remix is better in five tests with disk reading, encoding, and RAM speed all notably better. LPIA Ubuntu Netbook Remix appears better with write performance and encryption. It's not clear, given the test results for LPIA, whether the architecture has an advantage for end users (apart from the power advantage — see Chapter 3), and it's questionable if the results are sufficient evidence to continue maintaining support of LPIA by the Ubuntu Mobile team. Having said this, however, LPIA was never meant to be run with Ubuntu Netbook Remix so these results are in no way conclusive.

Chapter 9: Testing and Usability

At UDS Karmic in Barcelona, it was decided to drop official support for the LPIA architecture. Ubuntu MID will now be based on Mer upstream. Mer is a Linux operating system that's built on a thin base of Ubuntu Jaunty combined with the best open source elements of Nokia's Maemo platform. LPIA will be maintained by the Ubuntu community.

PBuilder for Automating the Testing of Packages

You can use pbuilder (see Chapter 5) to automate the testing of packages. It has a feature that allows hooks to be used; these hooks can try to install packages inside the chroot, run them, and even try to uninstall them. Some known tests include:

- An automatic install-remove-upgrade-remove-install-purge-upgrade-purge test-suite (distributed as an example called B91dpkg-i and found in the /usr/share/doc/pbuilder/examples directory). There is also a check that everything installs called execute_installtest.sh found in the same directory.

- Automatically running lintian (distributed as an example in /usr/share/doc/pbuilder/examples/B90lintian)

To use this test suite, hooks need to be created. Create a directory, as follows:

```
/var/cache/pbuilder/hooks
```

Add, for example, the B91dpkg-i script to this directory.

It is also possible to specify on the command line --hookdir /usr/share/doc/pbuilder/examples *to include all the example hooks.*

Once these hooks are established, test scripts can be run. These test scripts are placed in the source code for the package being tested, in the folder debian/pbuilder-test/NN_name (where NN is a number), following the run-parts standard for filenames.

Run-parts (from the debianutils *package) is used to execute a number of scripts or programs that are found in one directory. The "run-parts" program requires these script filenames to consist entirely of letters, digits, underscores, or hyphens.*

Example scripts that are used with pbuilder-test can be found in /usr/share/doc/pbuilder/examples/pbuilder-test.

An example of such a test script is test 004_ldd:

```bash
#!/bin/bash
# simple test to see that the program is installed in the correct
# place and looks sane through ldd.
ldd /usr/bin/program
```

The command ldd *lists the dynamic dependencies of executable files or shared objects.*

Chapter 9: Testing and Usability

Other Useful Linux Performance Testing Tools

Various tools already exist on Ubuntu which can help to optimize performance.

ps

This is a tool that allows you to view the processes that are running on the mobile device. If you want to search for a specific process, the following is a useful command:

```
ps aux | grep <something>
```

top

Top is a tool that enables you to keep track of processes; it lists processes that are ordered by CPU or memory usage. When put together with the `at` command, it is possible to schedule top to run at a particular time. You can achieve this by writing a bash script that runs top one minute later like this:

```
$ cat ./app-test.at
TERM=linux top -b -n 1 >/tmp/top-report.txt
$ at -f ./app-test.at now+1minutes
```

The application under test can then be started and its resource usage observed.

time

This tool measures execution times. It takes a command as an argument. It executes and then shows a brief statistic about the amount of time that the process spends. This can help you to find out if a command takes more time than expected. Of course, this is especially useful if you are dealing with a program that you have developed because you can modify the sources and then compile them again to compare results.

Time is used this way:

```
$ time ./pumd.py
real    1m41.365s
user    0m0.000s
sys     0m0.004s
```

procinfo

The load average is a measurement of the active processes running at any time. It is the sum of the run queue length and the number of jobs that are currently running on the CPU. The results are calculated using a program called procinfo. Here's an example:

```
$ sudo apt-get install procinfo
$ procinfo
Memory:         Total       Used        Free        Buffers
RAM:            1017380     966680      50700       11980
Swap:           3071992     118784      2953208

Bootup: Mon Jun  1 20:15:21 2009    Load average: 0.53 0.54 0.41 2/274 1186
```

201

Chapter 9: Testing and Usability

free

This application displays the amount of free memory and the amount of memory that is used by a system:

```
$ free
                total       used       free     shared    buffers     cached
Mem:          1017380     983448      33932          0      13084     207020
-/+ buffers/cache:         763344     254036
Swap:         3071992     120572    2951420
```

memstat

This application lists all the processes, executables, and shared libraries that are using up virtual memory:

```
$ memstat
```

which gives output like the following:

```
1032k: PID   5166 (/usr/lib/libindicate.so.1.0.0)
272k: PID    5178 (/lib/tls/i686/cmov/libnss_files-2.9.so)
736k: PID    5249 (/usr/lib/libltdl.so.7.2.0)
709848k: PID 5251 (/usr/share/fonts/truetype/ttf-dejavu/DejaVuSans.ttf)
43176k: PID  5256 (/usr/lib/libgvfscommon.so.0.0.0)
```

Look at the memory usage by the DejaVuSans font!

memcheck and Valgrind

Valgrind is a suite of tools which is used for debugging and profiling programs. It can be installed by using apt:

```
$ sudo apt-get install valgrind
```

Next, start the program to verify that the system is under the control of memcheck:

```
$ G_SLICE=always-malloc G_DEBUG=gc-friendly  valgrind -v --tool=memcheck
--leak-check=full --num-callers=40 --log-file=valgrind.log <application_name>
```

This command can take a while to run.

The resulting valgrind.log file displays an error summary :

```
==2753== ERROR SUMMARY: 0 errors from 0 contexts (suppressed: 323 from 4)
```

as well as a leak summary:

```
==2753== LEAK SUMMARY:
==2753==    definitely lost: 13,891 bytes in 43 blocks.
==2753==    indirectly lost: 24,510 bytes in 1,216 blocks.
==2753==      possibly lost: 1,128 bytes in 31 blocks.
==2753==    still reachable: 9,878,411 bytes in 56,308 blocks.
==2753==         suppressed: 0 bytes in 0 blocks.
```

Chapter 9: Testing and Usability

Valgrind is an excellent development tool for creating highly efficient software.

Latencytop

The Intel Open Source Technology Center released Latencytop, a tool aimed at identifying where in the system latency is happening. It is available in the Ubuntu Universe repository and can be installed using apt:

```
$ sudo apt-get install latencytop
```

First, run the following:

```
$ sudo latencytop
```

The results appear in a table with the Firefox process selected:

```
Cause                                               Maximum       Percentage
Scheduler: waiting for cpu                         12.7 msec           5.2 %
Waiting for event (poll)                            5.0 msec          64.4 %
Userspace lock contention                           5.0 msec          29.1 %
Waiting for event (select)                          4.6 msec           0.3 %
Reading from a pipe                                 2.7 msec           0.0 %
Waiting for a process to die                        2.1 msec           0.0 %
opening cdrom device                                1.5 msec           0.4 %
Executing raw SCSI command                          1.1 msec           0.4 %
acpi_ec_wait acpi_ec_transaction_unlocked acpi_ec_  1.0 msec           0.0 %

Process firefox (6030)                      Total: 294.5 msec
Scheduler: waiting for cpu                          9.1 msec          52.3 %
Userspace lock contention                           5.0 msec          25.5 %
Waiting for event (poll)                            4.8 msec          22.2 %

   evolution-data-    evolution-excha    firefox    gnome-terminal    gnome-pty-helper
```

Latencytop reads from the /proc/latency_stats file, and displays the largest latencies during the last 30 seconds. In this case, finding the Userspace lock contention seems a good place to try improving performance.

Testing Strategies

With many tools available for testing, the question of creating a testing strategy becomes important. For the Ubuntu MID distribution itself, quality assurance tests are divided into three groups: Basic, Advanced, and Compliance.

Basic

Basic feature testing covers the features that the end user of the software will see on a daily basis. Defects in this area of functionality will be noticed immediately by the user. It is vital that this area of testing be performed as often as possible; as such, it is a good candidate for automation. Applications that fit into this category include:

- Midbrowser
- RSS reader
- Clock
- Calculator
- PIM
- Remote desktop client
- E-book reader
- Notepad

Advanced

Advanced cases require either more system knowledge, or access to resources that not everyone will have, such as proprietary codecs.

Compliance

Compliance cases cover the agreed upon statistics for the system. There are some checklists that serve as a guide in this category (The following checklists are shown as examples only; the actual targets will vary according to a term of work and will be specific to each OEM):

Performance Tests (Excludes BIOS Times)

- With platform in Hibernate state, the time from the user pressing the Power On button to the desktop UI being fully responsive ≤ 16 sec
- Time from the user pressing the Sleep button to the platform going to S3 state (not just display going blank) ≤ 7 sec
- With platform in S3 state, time from the user pressing the Wake-Up button to the desktop UI being fully responsive ≤ 7 sec
- When the platform is cold booted, the time from the user pressing the Power On button to the desktop UI being fully responsive ≤ 37 sec

Footprint

- Platform memory 192MB
- Use a maximum of 2000MB

Localization

- Support English (US)

Chapter 9: Testing and Usability

Documentation

- ❏ Deliver updated documentation
- ❏ Licensing terms for all documentation
- ❏ List of any proprietary documentation that cannot be made public
- ❏ Licensing terms for all components
- ❏ List of known non–open source components that will be included

Bug Reporting

It is always important to file bug reports immediately when something does not go as planned.

If You Find a Bug . . .

It is always a good practice to submit bug reports. Bug reports should be submitted to `https://bugs.edge.launchpad.net/ubuntu-mobile/`.

When submitting bug reports, there is a template to follow, which gives developers the best chance of being able to reproduce — and hopefully — squash the bug.

The template looks like this:

```
Build Version/Date:
Environment used for testing:
Summary:
Steps to Reproduce:
Expected result:
Actual result:
```

The following is an example of a good bug report (which you'll find at `https://bugs.launchpad.net/ubuntu-mobile/+bug/218850`):

```
<Build: UME Daily 20081404>
<Environment: Xephry>
Summary:
When opening an office document the "Choose the file to Visualize" dialogue remains
persistent over the top of the document after it is loaded. Blocks all document
content.
Steps to Reproduce:
1) Launch Document Viewer
2) Select document (ppt, doc, pdf, etc.)
3) Open
Expected result:
Document is opened and content is displayed without error or corruption
Actual result:
 Viewer selection dialogue remains in foreground after document is opened
```

Chapter 9: Testing and Usability

Filing a Bug Report Automatically

The easiest way to report a bug is to use the Apport's application. When an application crashes, Apport starts automatically and opens an appropriate bug report for you to complete in Launchpad.

If you want to report a bug for an application that is running and responding, you can access the same Apport functionality from the Help menu via Help ➪ Report a problem. This method is preferable to filing a bug report directly at the Launchpad website because of the debugging information it attaches (including user interface bugs).

Reporting a Bug from the Command Line

To report a bug from the command line, use the following:

```
$ubuntu-bug <PACKAGE_NAME>
```

To file a bug report against a known package, and automatically include all the useful debug information, use the following:

```
ubuntu-bug PID
```

To file a bug report against a running program with a known Process ID (PID) (see System ➪ Administration ➪ System Monitor), use the following:

```
ubuntu-bug linux
```

This will file a bug report on the Linux kernel.

Summary

One advantage that Linux has over proprietary systems is the number of people who are prepared to help test and submit bug reports. Everyone can help make Ubuntu better by submitting bug reports. Setting up a good testing strategy increases the chances of finding bugs and contributing to the improvement of Ubuntu, and in turn, contributing to the wider open source ecosystem that surrounds it.

10
Tips and Tricks

This chapter contains a number of scripts, tips, and hints for using and enhancing a device running Ubuntu Mobile.

Improving Boot Speed

This section describes how to improve and increase boot speed on an Ubuntu Jaunty device. This is an area which is under very heavy development as the ambitious target boot speed for Ubuntu 10.04 (karmic+1) is 10 seconds.

> *karmic+1 refers to the release of Ubuntu after the karmic 9.10 release. The adjective which describes this release is, at the time of writing, undecided.*

For the karmic release, the target boot speed is expected to be somewhere between the current speed (28s on Jaunty) and the ambitious new target. A consequence of this (which was discussed at UDS Barcelona) is that both the initramfs and the X Server are likely to see a lot of work in the coming months, so the techniques described below may not be the best ways to reduce boot speed. Please refer to the ubuntu-boot mailing list for the latest information.

Hard Coding Modules

It is possible to list modules so that they are called when the initramfs is called. Doing this means that *only* the modules that the specific device needs to boot are loaded.

> *initramfs is a small ram-based initial root filesystem which the kernel runs as its first program as the device boots.*

Chapter 10: Tips and Tricks

To do this, put the necessary modules into

```
$ /etc/initramfs-tools/modules
```

and run

```
$ sudo update-initramfs -u
```

It is not necessary to change the /etc/initramfs-tools/initramfs.conf file (for example, if the `modules` *option still has the value* `most` *in the .conf script) because this is concerned with which modules are copied into the initramfs, not with which modules are probed.*

An example list might look like this:

```
loop
squashfs
```

The file

```
$ /etc/initramfs-tools/modules
```

is looped by the `load_modules` function from

```
$ /usr/share/initramfs-tools/init
```

Creating a /tmp That Is Half the Size of Physical RAM

You can add the following to /etc/fstab:

```
tmpfs /tmp tmpfs nodev,nosuid 0 0
```

This creates a /tmp that is half the size of physical RAM, and anything stored in /tmp goes to RAM. In practice, much of it ends up being committed to swap, but it means that it gets written to disk only when the space in /tmp exceeds the cache buffers (instead of immediately). This can speed up operations on temporary files for machines with slower access to secondary storage (not, however, on SSD).

In reality, this can improve the performance of most hard drives with speeds of 7200, 5400, or even slower.

Of course, the downside is that you want to clean up /tmp when hibernating the device, especially if the device is short on RAM anyway. Be careful!

Energy Tips

For product designers, an understanding of the factors affecting battery life is vitally important for managing product performance. End users need as much battery life as you can give them.

Chapter 10: Tips and Tricks

Recharging Correctly

As a user, try to recharge the device without interruptions. Also, when you first use the device, run the battery down to zero and then fully recharge. This calibrates the battery.

Laptop Mode

When laptop mode is enabled, the kernel will try to be smart about when to do writes in order to give the disk as much time as possible in a low power state.

Laptop-mode-tools are installed by running the following:

```
$ sudo apt-get install laptop-mode-tools
```

To start laptop mode, run `sudo laptop_mode start`, which shows the following:

```
Laptop Mode Tools 1.45
Laptop mode enabled, active
/dev/sda:
 setting Advanced Power Management level to 0xfe (254)
/dev/sda:
 setting standby to 12 (1 minutes)
```

Running `laptop_mode` increases dirty_expire_centisecs and dirty_writeback_centisecs in /proc/sys/vm, which means pages will not be written to disk as often. It also changes the ratio of dirty background writebacks, which taken together, result in rapid bursts of disk activity and longer dormant states.

The configuration file for laptop-mode is

```
/etc/laptop-mode/laptop-mode.conf
```

A tool (which comes with laptop-mode-tools) called lm-profiler will monitor the device disk usage and running network services and it will suggest disabling unneeded services. Running it gives an output like the following:

```
$ sudo lm-profiler

Write accesses at 445/600 in lm-profiler run: gconfd-2

Read accesses at 465/600 in lm-profiler run: hald-addon-stor

Read accesses at 526/600 in lm-profiler run: gconfd-2

Read accesses at 535/600 in lm-profiler run: chrome

Profiling run completed.


Program:      "anacron"
```

Chapter 10: Tips and Tricks

```
Reason:         standard recommendation (program may not be running)

Init script: /etc/init.d/anacron (GUESSED)

Do you want to disable this service in battery mode? [y/N]: Y
Program:        "cron"

Reason:         standard recommendation (program may not be running)

Init script: /etc/init.d/cron (GUESSED)

Do you want to disable this service in battery mode? [y/N]:Y

Program:        "atd"

Reason:         standard recommendation (program may not be running)

Init script: /etc/init.d/atd (GUESSED)

Do you want to disable this service in battery mode? [y/N]: y
```

and so on as the application "guesses" at services to disable. Choosing to disable some services when in laptop mode has a very positive effect on battery life.

Getting to Know the Battery on a Device

Run the following:

```
$ cat /proc/acpi/battery/BAT0/info
```

The result shows the design capacity (and the last full capacity of the battery). This command can be very useful when identifying battery errors:

```
present:                 yes
design capacity:         28800 mWh
last full capacity:      22240 mWh
battery technology:      rechargeable
design voltage:          14400 mV
design capacity warning: 1112 mWh
design capacity low:     200 mWh
capacity granularity 1:  1 mWh
capacity granularity 2:  1 mWh
model number:            92P1163
serial number:            2335
battery type:            LION
OEM info:                SANYO
```

The battery is a SANYO LION, which is starting to show some signs of deterioration (look at the difference between the design and the last full capacity).

If you have compiled your own kernel for a device, then there is a chance that the /proc interface will not be available. In this case, you can use the command:

```
$ ls /sys/class/power_supply/BAT0/
```

Chapter 10: Tips and Tricks

CPUFREQ and Governors

The power that is consumed by a processor is related to the speed at which it is running — it's a performance versus battery life trade-off. A way to manage this is to enable CPUFREQ in the running kernel, along with the different controlling algorithms — that is, the governors that control its use.

> *Most cpu frequency scaling algorithms or drivers only allow the CPU to be set to one frequency. In order to offer dynamic frequency scaling, the cpufreq core must be able to tell these drivers of a "target frequency." This is done by using cpufreq governors.*

Available governors can be listed by using this command:

```
$ cat /sys/devices/system/cpu/cpu0/cpufreq/scaling_available_governors
The output looks like:
userspace ondemand powersave conservative performance
```

The governor currently in use is shown by:

```
$ cat /sys/devices/system/cpu/cpu0/cpufreq/scaling_governor
```

which outputs:

```
ondemand
```

The chosen governor is loaded automatically from the init script /etc/init.d/powernowd. You can add options by modifying the file /etc/default/powernowd; for example, you can add the option `-m 3`, which puts powernowd into LEAPS mode, causing it to jump immediately to the highest frequency if usage is above 80 percent or to the lowest frequency if usage is below 20 percent.

Use Power Management Settings on Disks

To see if your disks support the setting of power management, you can run the following (substituting your device mount point):

```
$ sudo hdparm -i /dev/mapper/lawrence-root
```

If your hard drive supports AdvancedPM, you can use `hdparm` to tell the disk to go into power savings mode after an elapsed period of idle time. The relevant options for `hdparm` are:

- `-B` — Sets the Advanced Power Management setting (1–255)
- `-S` — Sets standby (spindown) timeout
- `-y` — Puts the IDE drive in standby mode
- `-Y` — Puts the IDE drive to sleep

Chapter 10: Tips and Tricks

For example, to put the disk into the most aggressive power savings mode after 60 seconds of idle time, you could use the following:

```
$ hdparm -B 1 -S 12 /dev/mapper/lawrence-root
```

Disabling atime

A significant downside to `atime` (a timestamp of file access required under the POSIX standard) is that every time a file is accessed, the kernel has to write a new timestamp to the disk. These disk writes will keep the disk as well as the link to the disk busy, and that costs both performance and power. `atime` can be disabled with the following:

```
$ sudo mount -o remount,noatime  /
```

Another approach that was added in the 2.6.20 kernel is to use the `relatime` mount option. If this flag is set, access times are updated only if they are earlier than the modification time. This change allows utilities to see if the current version of a file has been read, but still cuts down significantly on `atime` updates. `relatime` can be enabled with the following:

```
$ sudo mount -o remount,relatime /
```

Turning Off Background Services

Background services are listed by default in xdg. xdg defines a method for automatically starting applications during the startup of a desktop environment and after mounting a removable medium. For example:

```
# cd /etc/xdg/autostart/
# ls -la
total 76
drwxr-xr-x 2 root root 4096 2008-07-06 20:30 .
drwxr-xr-x 6 root root 4096 2008-07-06 20:27 ..
-rw-r-r- 1 root root 2676 2008-03-28 07:07 bluetooth-applet.desktop
-rw-r-r- 1 root root 289 2008-04-21 12:23 evolution-alarm-notify.desktop
-rw-r-r- 1 root root 5040 2008-04-15 11:42 gnome-at-session.desktop
-rw-r-r- 1 root root 6079 2008-04-10 14:12 gnome-power-manager.desktop
-rw-r-r- 1 root root 7214 2008-04-15 05:39 gnome-volume-manager.desktop
-rw-r-r- 1 root root 293 2008-04-08 22:06 jockey-gtk.desktop
-rw-r-r- 1 root root 374 2008-04-18 17:27 nm-applet.desktop
-rw-r-r- 1 root root 219 2008-04-06 22:24 pulseaudio-module-xsmp.desktop
-rw-r-r- 1 root root 5002 2008-04-21 12:44 redhat-print-applet.desktop
-rw-r-r- 1 root root 2204 2008-04-02 17:11 tracker-applet.desktop
-rw-r-r- 1 root root 1791 2008-04-02 17:11 trackerd.desktop
-rw-r-r- 1 root root 236 2008-04-04 18:34 update-notifier.desktop
-rw-r-r- 1 root root 2783 2008-02-13 07:51 user-dirs-update-gtk.desktop
```

If, for example, you do not wish to have tracker running, you can move it into another folder with the following:

Chapter 10: Tips and Tricks

```
# mkdir ../old
# mv tracker-applet.desktop trackerd.desktop ../old
```

The next time the desktop starts, tracker will not be started.

Adobe Flash

Using Flash on a battery powered, mobile device can be somewhat challenging. This is unfortunate as Flash is a popular software application and, indeed, was once used as a "proof-of-concept" Home area for the Ubuntu Hardy MID, as shown in Figure 10-1.

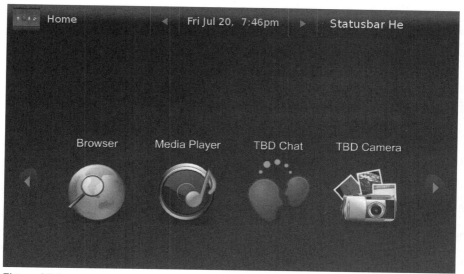

Figure 10-1

According to Adobe, Flash will always maximize the CPU if there are the cycles to spare. It does this because, for users to perceive animation as playing smoothly, it is important that each frame be displayed at the proper time. Even small inconsistencies in frame-to-frame timing can make animation appear jerky.

The problem with this is that a new model of scheduling called "dynamic tick" was introduced in version 2.6.21 of the Linux kernel, which allows the processor to be almost entirely shut down until the next time the user interacts with the computer. Such constant polling by Flash does not allow this to happen, which has a devastating effect upon power consumption.

Chapter 10: Tips and Tricks

CPU usage can be monitored using `top` or graphically using `gnome-system-monitor`, which is available through apt.

Although the CPU problem cannot be eliminated entirely without code updates from Adobe, some improvements were noticed during experiments running the Linux .tar.gz version from Adobe (not the Flash version packaged in the Ubuntu repositories):

```
$ cd /tmp
$ wget http://fpdownload.macromedia.com/get/flashplayer/current/install_flash_player_10_linux.tar.gz
$ tar -xvf install_flash_player_10_linux.tar.gz
$ cd  install_flash_player_10
$ sudo ../flashplayer-installer
```

This suggests that work is ongoing in this area.

There are persistent rumors of a Flash iPhone version, so this is definitely something to watch.

Configuring the Touchscreen

Touchscreens are more and more common in all kinds of devices, from Notebooks to MIDs to all-in-one PCs. For Ubuntu Jaunty, evtouch will go on to be maintained as the solution for the mobile team as it covers the broadest set of touchscreens.

Evtouch is a touchscreen driver for X. The driver is actually an evdev-driver, which supports events for moving in absolute coordinates, relative coordinates, and mouse buttons.

The graphical touchscreen calibration tool on Ubuntu MID is accessed through Preferences ⇨ Calibrate Touchscreen. If this does not work as expected on a device, it is possible to manually run the calibrate script. Run:

```
$ lshal | grep evtouch
```

This ensures that the device is using the evtouch driver rather than, say for example, the evdev one. If this command returns empty, do the following:

```
sudo apt-get install xserver-xorg-input-evtouch
```

Stop and then start the X Server, or reboot.

Then run the calibration touchscreen program:

```
# /usr/lib/xf86-input-evtouch/calibrate.sh
```

When this is completed, look in /etc/evtouch/config and note the answers.

Chapter 10: Tips and Tricks

On some devices, the results seem to be inverted. In this case, it is good to know that x0,y0 should be the bottom left of the screen. You might have to tweak these coordinates. Increase numbers to move the cursor to the right, or up, relative to the stylus. It is likely that if you have values over 20 or below −20 they are likely to be wrong.

Once these values have been worked out correctly, put them into a new hal configuration file /etc/hal/fdi/policy/touchscreen.fdi

A hardware abstraction layer (hal) is an abstraction layer, implemented in software, between the physical hardware of a device and the software that runs on the device.

This file looks like

```xml
<?xml version="1.0" encoding="UTF-8"?> <!-- -*- SGML -*- -->
<deviceinfo version="0.2">
  <device>
    <match key="info.product" contains="TouchScreen">
      <match key="info.capabilities" contains="input">
        <merge key="input.x11_driver" type="string">evtouch</merge>
        <merge key="input.x11_options.minx" type="string">59</merge>
        <merge key="input.x11_options.miny" type="string">82</merge>
        <merge key="input.x11_options.maxx" type="string">963</merge>
        <merge key="input.x11_options.maxy" type="string">989</merge>

        <!--bottom, left to right-->
        <merge key="input.x11_options.x0" type="string">-3</merge>
        <merge key="input.x11_options.x1" type="string">7</merge>
        <merge key="input.x11_options.x2" type="string">0</merge>

        <!--mid, left to right-->
        <merge key="input.x11_options.x3" type="string">-9</merge>
        <merge key="input.x11_options.x4" type="string">3</merge>
        <merge key="input.x11_options.x5" type="string">14</merge>

        <!--top, left to right-->
        <merge key="input.x11_options.x6" type="string">-15</merge>
        <merge key="input.x11_options.x7" type="string">2</merge>
        <merge key="input.x11_options.x8" type="string">21</merge>

        <!--bottom, left to right-->
        <merge key="input.x11_options.y0" type="string">5</merge>
        <merge key="input.x11_options.y1" type="string">0</merge>
        <merge key="input.x11_options.y2" type="string">0</merge>

        <!--mid, left to right-->
        <merge key="input.x11_options.y3" type="string">4</merge>
        <merge key="input.x11_options.y4" type="string">6</merge>
        <merge key="input.x11_options.y5" type="string">4</merge>

        <!--top, left to right-->
        <merge key="input.x11_options.y6" type="string">-2</merge>
        <merge key="input.x11_options.y7" type="string">-3</merge>
        <merge key="input.x11_options.y8" type="string">-4</merge>
```

Chapter 10: Tips and Tricks

```xml
                <merge key="input.x11_options.taptimer" type="string">30</merge>
                <merge key="input.x11_options.longtouchtimer" type="string">750</merge>
                <merge key="input.x11_options.longtouched_action" type="string">click</merge>
                <merge key="input.x11_options.longtouched_button" type="string">3</merge>
                <merge key="input.x11_options.oneandhalftap_button" type="string">2</merge>
                <merge key="input.x11_options.movelimit" type="string">1</merge>
                <merge key="input.x11_options.touched_drag" type="string">1</merge>
                <merge key="input.x11_options.maybetapped_action" type="string">click</merge>
                <merge key="input.x11_options.maybetapped_button" type="string">1</merge>
            </match>
        </match>
    </device>
</deviceinfo>
```

If the touchscreen calibrates but is not very accurate (examples of this might be dragging the finger or stylus across the screen and the cursor jumping about ¼ to ½ inch at a time), then try adding a value for the `movelimit` parameter (make it 5 or 10). This defines the step size the evtouch driver uses, which defaults to 30 pixels and this might be too large a value for the device.

> *Have a look at the /usr/share/hal/fdi/policy/10osvendor/50-* files for help with this and to create your own .fdi file.*

Restart hal:

```
# service hal restart
```

Next, restart X. Your touchscreen should now be working. If not, you need to make sure that *your* custom .fdi file (if you made one) is being used.

It is important to remember that the files in /etc/hal/fdi/* will have a higher priority than those from /usr/share/hal/fdi/; you may also need to check that X is picking up your changes. To do this, run the following:

```
$ UDI=$(hal-find-by-capability --capability input.mouse)
$ lshal -u $UDI
```

This gives you something like this:

```
udi = '/org/freedesktop/Hal/devices/usb_device_eef_1_noserial_if0_logicaldev_input'
  info.capabilities = {'input', 'input.mouse'} (string list)
  info.category = 'input' (string)
  info.parent =
'/org/freedesktop/Hal/devices/usb_device_eef_1_noserial_if0'  (string)
  info.product = 'eGalax Inc. Touch' (string)
  info.subsystem = 'input' (string)
```

Chapter 10: Tips and Tricks

```
      info.udi =
'/org/freedesktop/Hal/devices/usb_device_eef_1_noserial_if0_logicaldev_input'
  (string)
      input.device = '/dev/input/event1'  (string)
      input.originating_device =
'/org/freedesktop/Hal/devices/usb_device_eef_1_noserial_if0'  (string)
      input.product = 'eGalax Inc. Touch'  (string)
      input.x11_driver = 'evtouch'  (string)
      input.x11_options.longtouched_action = 'click'  (string)
      input.x11_options.longtouched_button = '3'  (string)
      input.x11_options.longtouchtimer = '750'  (string)
      input.x11_options.maxx = '1912'  (string)
      input.x11_options.maxy = '1989'  (string)
      input.x11_options.maybetapped_action = 'click'  (string)
      input.x11_options.maybetapped_button = '1'  (string)
      input.x11_options.minx = '112'  (string)
      input.x11_options.miny = '76'  (string)
      input.x11_options.movelimit = '10'  (string)
      input.x11_options.oneandhalftap_button = '2'  (string)
      input.x11_options.rotate = 'ccw'  (string)
      input.x11_options.swapy = true  (bool)
      input.x11_options.taptimer = '30'  (string)
      input.x11_options.touched_drag = '1'  (string)
      linux.device_file = '/dev/input/event1'  (string)
      linux.hotplug_type = 2  (0x2)  (int)
      linux.subsystem = 'input'  (string)
      linux.sysfs_path = '/sys/class/input/input1/event1'  (string)
```

Now you can see the options that hal has for the device. You need to be sure there is not a configuration for "Input Device" on the xorg.conf file in order to let hal configure the X file.

Watching Hard Disk Activity

An application called iotop watches I/O usage information output by the Linux kernel (it requires 2.6.20 or later) and displays a table of current I/O usage by processes or threads on the system.

The CONFIG_TASKSTATS and CONFIG_TASK_IO_ACCOUNTING options need to be enabled in your Linux kernel build configuration.

iotop displays columns for the I/O bandwidth that is read and written by each process/thread during the sampling period. It also displays the percentage of time the thread/process spent while swapping in and while waiting on I/O. In addition, the total I/O bandwidth that is read and written during the sampling period is displayed at the top of the interface.

Chapter 10: Tips and Tricks

First, install the following:

```
sudo apt get install iotop
```

Next, run the following:

```
$ iotop
```

To watch a specific process, pass the PID to the application like this:

```
$ iotop -p 7279
```

Summary

The tips and tricks come from the Ubuntu Mobile team, IRC discussions, and e-mails from users. If you discover something useful that is not covered here, add it directly to the FAQ on the Ubuntu Mobile wiki. Thanks in advance for sharing!

11
Putting It All Together

This chapter walks you through the process of creating a custom distribution of Ubuntu Mobile, which could potentially be used by an OEM to go to market. It could just as easily be used by a "homebrewer" who wishes to resurrect an old, unused mobile device that's gathering dust.

Like beer making, no special equipment (apart from a laptop and Internet connection) is needed to create your image. Choosing a good beer recipe and following good brewing procedures can give you top quality beer even with basic equipment. The same principles apply to creating a custom image.

We draw from the previous chapters and show you how to use the techniques and skills that you learned in those chapters in order to undertake an example project — creating a customized Ubuntu Mobile image for Ubuntu karmic that can be installed on a target device.

Important Things to Consider

When creating an image for release, there are some key questions to ask — the answers to which will influence the content of the image and the default settings on the device.

Ubuntu is updated on a six-month schedule. Within this six-month release cycle there are standard, Long Term Support (LTS) desktop, LTS server, and point releases.

If what is required is the bleeding edge software, the latest *point* release should be used. However, it can make more sense (especially for OEMs) to build on an LTS release. A new LTS version is released every two years.

With the Long Term Support (LTS) versions, commercial support is available from Canonical. This is available for three years on the desktop and five years on the server. (24/7 commercial support for Ubuntu Mobile is available through Canonical's global support team and partners. See `http://www.ubuntu.com/support/paid` for more details.)

Chapter 11: Putting It All Together

Check If the Device Architecture Is Supported by Ubuntu

The appropriate Ubuntu architecture must be chosen based on the hardware architecture. The current ports, at the time of this writing, are armel, hppa, ia64, lpia, powerpc, and sparc. The main supported architectures are i386 and amd64.

Checking the Hardware

Hardware support can be an area of some concern. It is a good idea to always check with the bios supplier if it is up-to-date and that there is support for Linux.

A useful program is hwinfo:

```
$ sudo apt-get install hwinfo
```

It provides a detailed report of all hardware on the device.

> *The lshw program provides a subset of the information hwinfo presents. Also, the lspci program is handy when working with hardware as it prints detailed information about all the PCI buses.*

Other useful resources include the Hardware Compatibility Lists, or HCL. These list the hardware components, parts, and add-ons that are used in netbooks, MIDs, PDAs, mobile phones, media players, GPS devices, and "wearables." For example, to check webcam support for a device, go to http://tuxmobil.org/laptop_webcams_linux.html. Choose the target device if listed. Clicking the Acer Aspire One ZG5 link gives all types of information about the hardware on the device.

For example, this is the information given on the webcam:

```
uvcvideo: Found UVC 1.00 device Acer Crystal Eye webcam (064e:d101)
usbcore: registered new interface driver uvcvideo

filename:       /lib/modules/2.6.23.91w/usb/media/uvcvideo.ko
version:        SVN r215
license:        GPL
description:    USB Video Class driver
author:         Laurent Pinchart
srcversion:     9F1EDDCB5114CB7A45B4A74
..
depends:        videodev,v4l2-common,v4l1-compat,compat_ioctl32
vermagic:       2.6.23.91w SMP preempt mod_unload CORE2
parm:           quirks:Forced device quirks (uint)
parm:           trace:Trace level bitmask (uint)
```

This device is supported out-of-the-box by Ubuntu. An `apt-cache` search shows that the following applications are available in the repositories:

- **luvcview** — USB Video Class grabber
- **uvccapture** — USB UVC Video Class snapshot software

so these could perhaps be added to the seed.

Chapter 11: Putting It All Together

Fine-Tuning the Kernel

It may be necessary on a low–disk space device to remove all unnecessary drivers and modules from the kernel. It is also obviously a good idea to optimize processors (such as by enabling hyperthreading), manage memory in a coherent manner, and ensure that network parameters are correct.

The Atom processor on the target device is a new architecture, but based on older technologies. It's the first in-order x86 from Intel since the Pentium.

> *A processor receives instructions one by one and puts them in a pipeline and then executes them. In an in-order architecture, the instructions are executed in the order in which they arrive, whereas an out-of-order architecture is capable of changing the order in the pipeline.*

On the Atom, there is a long pipeline coupled to an in-order architecture — consequently hyperthreading can be effective and significantly increases performance. This needs to be enabled in both the kernel and the BIOS of the device itself.

> *CPU performance can be measured with htop and sysstat, which are available in repositories.*

To find out how to actually make such changes, read Chapter 8, "Kernel Fine-Tuning."

Defining Power Policies

DeviceKit-Power (see Chapter 3, "Power Management") is now integrated into the new Gnome-Power-Manager for karmic.

It is important to note that since Ubuntu Jaunty, powernow-k8 and acpi-cpufreq are no longer available as modules but are compiled into the kernel. This is important as these modules are used by projects such as Linux-PHC to *undervolt* a CPU to expand battery time and to reduce the CPU's temperature while not affecting performance.

As a result, to enable undervolting it is necessary to either compile a kernel with the stand-alone modules (this is at the expense of boot time) or to use the kernel from the Linux PHC PPA at `https://launchpad.net/~linux-phc/+archive/ppa`.

> *Linux-PHC utilizes the production tolerance of a CPU. CPUs have different production qualities so the vendor defines voltages for every CPU – so even those of low quality can be undervolted. Actual power performance results for undervolting on the Atom processor are mixed at present, so it is recommended to also optimize the system as fully as possible for power savings, suspending devices when possible (WLAN, USB, LAN).*

Is It an Embedded System?

Depending on its hardware type, the filesystem may need to be compacted. When creating embedded systems, every byte of the storage on the device is important, so compression is used everywhere possible.

Chapter 11: Putting It All Together

SquashFS is a read-only filesystem that compresses whole filesystems or single directories and then writes them to other devices/partitions or to ordinary files. These can then be mounted directly (if a device) or using a loopback device (if it is a file).

There is a trick to enable optimizations on a Solid State Drive (SSD) filesystem. Add elevator=noop to /boot/grub/menu.lst if using the original Grub, or add it to the configuration file in Grub2. See the following section for more information about Grub.

> elevator=noop *helps speed I/O reorder requests to the disk. This means that when the head moves across the disk, it can service those requests in an orderly, sequential manner, rather than going back and forth.*

Now that some basic questions have been answered, it is possible to think about what customizations will be required in the image. When working with an OEM, this will likely be agreed upon beforehand in a Terms of Work agreement. If you are "homebrewing," you will know what customizations that you want to make.

Customizing the User Interface

The default Ubuntu behavior can be heavily modified. Everything from the boot selector to the desktop theme can de changed. This section discusses some of the relevant customizations that can be made.

Boot Selector

For karmic, the default boot selector is Grub2. Grub2 is an improvement on the original Grub boot loader as it can be more easily customized. With the new design, other architectures are also supported. For now, however, it cannot be used as a bootloader for the ARM architecture. In this case. u-boot is used instead.

Grub2 now comes with a configuration file in /etc/default/grub, which contains information formerly contained in the old Grub menu.lst. The file looks like this:

```
# This file is sourced by update-grub, and its variables are propagated
# to its children in /etc/grub.d/

GRUB_DEFAULT=0
GRUB_TIMEOUT=4
GRUB_DISTRIBUTOR=`lsb_release -i -s 2> /dev/null || echo Debian`
GRUB_CMDLINE_LINUX_DEFAULT="quiet splash"

# Uncomment to disable graphical terminal (grub-pc only)
#GRUB_TERMINAL=console

# Uncomment if you don't want GRUB to pass "root=UUID=xxx" parameter to Linux
#GRUB_DISABLE_LINUX_UUID=true
```

In addition to the file /etc/default/grub, the folder /etc/grub.d/ contains other files that are read during the execution of update-grub or update-grub2 commands. The contents are then imported into /boot/grub/grub.cfg.

Chapter 11: Putting It All Together

These files are:

- **00_header**
- **05_debian_theme** — Set background and text colors, themes.
- **10_hurd**
- **10_linux**
- **20_memtest86+** — If the /boot/memtest86+.bin file exists, it is included as a menu item.
- **30_os-prober** — Searches for other OSs and includes them in the menu.
- **40_custom** — A template for adding custom menu entries, which will be inserted into grub.cfg upon execution of the update-grub2 command. This file, as well as any other custom file, must be made executable to allow importation into grub.cfg.

The 05_debian_theme file can then be edited to set a different background image, as shown in the following code:

```
# check for usable backgrounds
use_bg=false
if [ "$GRUB_TERMINAL" = "gfxterm" ]; then
   for i in {/boot/grub,/usr/share/images/desktop-base}/Windbuchencom.{png,tga}; do
      if bg=`convert_system_path_to_grub_path $i`; then
         case ${bg} in
            *.png)          reader=png;;
            *.tga)          reader=tga;;
            *.jpg|*.jpeg)   reader=jpeg;;
         Esac
         if test -e /boot/grub/${reader}.mod; then
            echo "Found Debian background: `basename ${bg}`" >&2
            use_bg=true
            break
         fi
      fi
   done
fi
```

After changing the Windbuchencom image for the custom one, run the grub-install command to write the changes to the Grub2 configuration.

Display Manager

Display Manager is a program that can manage several X displays/sessions. The X display/session can be started by an X Server on a local or remote host. If the X Server is on a remote host, it requests a session to the Display Manager through the XDMCP, X Display Manager Control Protocol.

The Display Manager provides a service that prompts for login name and password, authenticating the user and running a session. The session may contain several programs, but mainly from a user's point of view, this will be a window manager.

Several Display Managers are available to choose from, but for this distribution, GDM is the choice.

Chapter 11: Putting It All Together

GDM

GDM is the Display Manager from the GNOME project. It is highly customizable and has an easy configuration and pre-configuration system.

Pre-Configuring GDM

The default `gdm` behavior can be changed. The config file follows the `ini` file format. The configuration includes several sections: `daemon`, `security`, `xdmcp`, `gui`, `greeter`, `chooser`, `debug`, and `servers`. The configuration file is found in /etc/gdm/gdm.conf-custom.

The following code shows how to enable the onscreen keyboard:

```
# gdm.conf-custom
# gdm
[daemon]
AddGtkModules=true
GtkModulesList=gail:atk-bridge:/usr/lib/gtk-
2.0/modules/libdwellmouselistener:/usr/lib/gtk-2.0/modules/libkeymouselistener
[security]
[xdmcp]
[gui]
[greeter]
SoundOnLogin=false
[chooser]
[debug]
[servers]
```

GDM can also be themed. This is explained in Chapter 7, "Theming."

Setting the Default Ubuntu, XFCE, and Hildon Behaviors

Gtk-based desktops use `GConf` as the default database for configurations. The following file is an example of how to set a default configuration for `gconf`. This file is usually included in a src/debian directory of a configuration package:

```
# <package-name>.gconf-defaults
/desktop/gnome/interface/gtk_theme brdesktop
/desktop/gnome/background/picture_filename /usr/share/wallpapers/brdesktop03.jpg
/desktop/gnome/volume_manager/autophoto_command "gthumb — import-photos %h"
/desktop/gnome/volume_manager/autoburn_data_cd_command brasero -d
/apps/gnome-session/options/show_splash_screen true
/apps/nautilus/preferences/always_use_browser true
/apps/gksu/sudo-mode false
/desktop/gnome/applications/main-menu/file-area/user_specified_apps
[iceweasel.desktop,ooo-writer.desktop,listen.desktop,gthumb.desktop,nautilus-home.desktop]
/apps/nautilus/preferences/preview_sound never
/desktop/gnome/applications/main-menu/system_monitor baobab.desktop
```

Fine-Tuning the Build Process

The following two sections are included here because both help considerably with the time it takes to build a final CD. It is better to play around with configurations and seeds than it is to be waiting for packages.

Setting Up a Repository

Setting up a Debian repository was covered in Chapter 5. To summarize, adding a debian package to a local repository is done as follows

```
$ reprepro includedeb <some-debian-package>
```

There may also be instances when a developer needs to add debian packages to a remote server (a PPA can also be used for this but, as mentioned previously, a PPA can take a long time to make packages available). To add a debian package to a remote server, install dupload:

```
$ sudo apt-get install dupload
```

Add a configuration in /etc/dupload.conf:

```
$cfg{'pumd} = {
        fqdn => "professional-ubuntu-mobile-development.com",
        incoming => "/pub/UploadQueue/",
};
```

This sets the fully qualified domain name to `professional-ubuntu-mobile-development.com` and the incoming folder on the server as /pub/UploadQueue/. Now it is possible to upload to this server using

```
$ dupload --to pumd <some-debian-package>.changes
```

Now with reprepro installed on the server, the package can be added to the repository. This can be scripted and the script added to a cron job. For more on this approach, go to www.debian-administration.org/articles/286.

Caching Packages with approx

If the Internet connection is limited, a caching system is very helpful. A number of tools can be used to cache packages such as approx, apt-cacher, apt-cacher-ng, and apt-proxy. Choosing one is a matter of preference but we have had good results from approx.

This is installed by running the following:

```
$ sudo apt-get install approx
```

A configuration file is located in /etc/approx/approx.conf. You can add the repository to it. Here is an example:

```
packages        http://professional-ubuntu-mobile-development.com/
```

In /etc/apt/source.list, add a line such as the following:

```
deb http://professional-ubuntu-mobile-development.com/packages/ubuntu karmic main
```

Creating a Default Ubuntu Image

Now is the time to think about actually creating the image. This image should be installable on a device and to accomplish this, an installer is necessary.

The techniques that are demonstrated in the rest of the chapter are useful not only when working with mobile devices, but also can be easily adapted to create installation images for desktops and servers.

Choosing Which Type of Installer to Use

There are two different types of installer; determining which one to use will depend to some degree on the target audience for the device and what type of image that you want to create.

When to Use Debian-Installer (Ubuntu Alternate Image)

The alternate installer is often used for more advanced installations as it provides options that are not available with the regular desktop image. This might include passing options to kernel modules or forcing a static network configuration. Let's consider another example: In Chapter 6 a user's home directory was encrypted. If this needs to be extended to encrypting a whole drive on a device, then this option is available only in the alternate image.

The alternate should also be used if the installation needs to be automated across a number of devices or if the device itself has a low memory footprint (less than 256MB RAM).

When to Use Ubiquity (Ubuntu Desktop Image)

Ubiquity is a graphical installer for Ubuntu, written in Python, which uses the debian-installer (d-i) as a backend for many of its functions. Ubiquity was built because the original interface to d-i was seen by some as fairly crude. This is because d-i uses a system called debconf to build its user interface, and because this interface depends on responses to questions it can lack a certain flexibility. Ubiquity is different as its primary focus is on ease of use for new users, and so it uses a filter called debconffilter, which sits between d-i code and the debconf front end, intercepting the debconf protocol, and transforming relevant debconf protocol messages into events that can be handled in much the same way as UI events from widget libraries like GTK+ or Qt.

> More info about the debian-installer can be found at `http://d-i.alioth.debian.org/doc/talks/debconf6/paper/`.

Getting Started on the Image: Preparing the Environment

There are many ways to build Ubuntu images. A variety of existing tools attempt to automate this task — such as remastersys, uck (Ubuntu Customization Kit), and methods that manually edit the default Ubuntu images. This section will describe the official Ubuntu way to create images.

Chapter 11: Putting It All Together

Creating an image from scratch involves four projects:

- cdimage
- debian-cd
- germinate
- britney

Each of these is explained in this section.

cdimage is the set of scripts that drives the whole process and handles synchronization with a local repository. It generates the debian-cd task lists from seeds and calls debian-cd to burn the finished ISO.

The first step for creating an ISO is to lay out a sane directory structure:

```
$ mkdir -p /home/<username>/devel/ubuntu
```

Change into the recently created directory. Next, make a branch of our ubuntu-cdimage scripts:

```
$ bzr branch bzr+ssh://bazaar.launchpad.net/~rclbelem/ubuntu-cdimage/mainline/ubuntu-cdimage
```

Also make a branch of the debian-cd with changes for Ubuntu:

```
$ bzr branch bzr+ssh://bazaar.launchpad.net/%7Eubuntu-cdimage/debian-cd/ubuntu/debian-cd
```

Create a symbolic link between the two:

```
$ ln -s debian-cd ubuntu-cdimage/
```

Britney is an updater script that intelligently moves packages (originally in Debian from the unstable repositories to the testing repositories), and the intelligent part comes from the fact that the destination for the packages should always be fully installable and close to being a release candidate. Branch this:

```
$ bzr branch http://people.ubuntu.com/%7Ecjwatson/bzr/britney/cdimage/ britney
```

Make a symbolic link from britney to ubuntu-cdimage:

```
$ ln -s britney ubuntu-cdimage/
```

Branch germinate, as follows:

```
$ bzr branch bzr+ssh://bazaar.launchpad.net/%7Ecjwatson/germinate/mainline/germinate
```

Link it to ubuntu-cdimage:

```
$ ln -s germinate ubuntu-cdimage/
```

Chapter 11: Putting It All Together

Now install debian-cd, debmirror, and procmail. We will not use debian-cd itself, but installing it (and the other packages mentioned) means that all the necessary dependencies are satisfied.

```
$ sudo apt-get install debian-cd debmirror procmail zsync
```

Install germinate to install its dependencies, too.

```
$ sudo apt-get install germinate
```

Next, install a package on which britney depends:

```
$ sudo apt-get install libapt-pkg-dev
```

Create the base directory where the gpg files will be placed:

```
$ mkdir -p /home/<username>/devel/ubuntu/ubuntu-cdimage/secret/dot-gnupg
```

Now create a new gpg key. Answer the questions as they are asked during creation but make sure that the password field is empty (just hit Enter at the password prompt). The scripts in cdimage run in batch mode and so will fail if a password is set:

```
$ gpg — homedir /home/<username>/devel/ubuntu/ubuntu-cdimage/secret/dot-gnupg —
gen-key — force-v4-certs
```

Edit the recently created key and set it as the main key:

```
$ gpg -- homedir /home/<username>/devel/ubuntu/ubuntu-cdimage/secret/dot-gnupg —
edit-key <keyid>
```

This is done by doing the following

```
Command>1
Command>primary
```

Now import the Ubuntu keyring – this adds all the developer keys:

```
$ gpg — no-default-keyring — keyring /home/<username>/devel/ubuntu/ubuntu-cdimage/
secret/dot-gnupg/pubring.gpg — import /usr/share/keyrings/ubuntu-archive-keyring.gpg
```

> If the mirror will be hosting "feisty" repositories, the following step is needed (otherwise, skip it):
>
> ```
> gpg-no-default-keyring-keyring /home/<username>/devel/ubuntu/
> mirror-keyring/trustedkeys.gpg-import /usr/share/keyrings/
> ubuntu-archive-keyring.gpg
> ```

Pick a mirror from the list at https://launchpad.net/ubuntu/+archivemirrors that is close to your physical location and that also has rsync available, and then run the following command to create the local mirror:

Chapter 11: Putting It All Together

```
$ GNUPGHOME=/home/<username>/devel/ubuntu/ubuntu-cdimage/secret/dot-gnupg debmirror
-a i386 -s main,restricted,main/debian-installer -e rsync -r:ubuntu -h archive.
ubuntu.com -d karmic /home/<username>/devel/ubuntu/ubuntu-cdimage/ftp
```

The preceding command will take around 15GB of the disk space and will take a long time to download!

After following the steps, the environment is ready. It is time to build a default Ubuntu image.

Finally, Building the Default ISO

Now everything is ready to build the image. To do this, create a directory called seeds and download the seed file required:

```
$ mkdir seeds
$ cd seeds
$ bzr branch http://bazaar.launchpad.net/~ubuntu-core-dev/ubuntu-seeds/platform.karmic/
$ bzr branch http://bazaar.launchpad.net/~ubuntu-core-dev/ubuntu-seeds/ubuntu.karmic/
```

Edit the config file in ubuntu-cdimage/etc:

```
$ cd ./ubuntu-cdimage/etc/

$ vi config
```

The finished file looks like this:

```
#! /bin/sh
# Settings for building Ubuntu CD images.  The build procedure also involves
# syncing a local Ubuntu mirror; see etc/anonftpsync for settings affecting what
# is mirrored and where.
export LC_ALL=C
export CDIMAGE_ROOT="${CDIMAGE_ROOT:-/home/<username>/devel/ubuntu/ubuntu-cdimage}"
if [ -z $PROJECT ]; then
    PROJECT="${PROJECT:-ubuntu}"
fi
if [ -z $CAPPROJECT ]; then
    CAPPROJECT="${CAPPROJECT:-Ubuntu}"
fi
if [ -z $DIST ]; then
    DIST="${DIST:-karmic}"
fi
if [ -z $ARCHES ]; then

    case $DIST in
    warty|hoary|breezy|dapper|edgy)
    ARCHES="${ARCHES:-amd64 i386 powerpc}"
    ;;
    feisty|gutsy|hardy|intrepid|jaunty|karmic)
    ARCHES="${ARCHES:-i386}"
    ;;
    esac
fi
```

Chapter 11: Putting It All Together

```
# Do not update the local mirror
CDIMAGE_NOSYNC=1
LOCAL_SEEDS=file:///home/<username>/devel/ubuntu/seeds/ CDIMAGE_ROOT=`pwd`
CDIMAGE_INSTALL=1
CPUARCHES="$(echo "$ARCHES" | xargs -n1 | sed 's/+.*//' | sort -u | xargs)"
GNUPG_DIR="$CDIMAGE_ROOT/secret/dot-gnupg"
SIGNING_KEYID=1421BE56
# Mirror info
MIRROR=

MIRRORDIR=
# Hosts that need to be notified when the build is done.  Third-party users
# will want to keep this variable empty.
# The "async" mirrors will be notified asynchronously, i.e. we won't wait for
# them to respond.
TRIGGER_MIRRORS=
TRIGGER_MIRRORS_ASYNC=
export TRIGGER_MIRRORS
export TRIGGER_MIRRORS_ASYNC

# Some older versions of debootstrap must *think* they're running as root,
# even though that's not really needed here.  If you have that problem, make
# sure you have fakeroot installed and uncomment this variable.
DEBOOTSTRAPROOT=fakeroot
export DEBOOTSTRAPROOT
PATH="$CDIMAGE_ROOT/bin:$PATH"
export PATH
umask 002
```

Set the $PATH to the ubuntu-cdimage `bin` directory:

```
$ export PATH=/home/<username>/devel/ubuntu/ubuntu-cdimage/bin/:$PATH
```

and build the .iso:

```
$ CDIMAGE_ROOT=/home/<username>/devel/ubuntu/ubuntu-cdimage CDIMAGE_NOSOURCE=1
DIST=karmic ARCHES=i386 CDIMAGE_NOSYNC=1 for-project ubuntu cron.daily
```

This will then build the image.

Building a Customized Ubuntu Image

It is now possible to build a customized Ubuntu image (with whatever selection of packages and configuration is appropriate to your target users). The only limits to what can be achieved with this are your imagination and the time required to try out the different combinations of applications and configurations. For this, it is necessary to understand how seeds and the germinate program work.

There are many packages in the Ubuntu archive, and Canonical needs to support the ones it selects to enter into a release for at least 18 months. Keeping track of all these packages would become unmanageable eventually, and this is important as Canonical is offering commercial support for some but not all of these packages. Because of this, the Ubuntu developers created a system of seeds that are "germinated" into a full list of packages required for a specific task. Everything built from the Ubuntu archive has a corresponding set of seeds listing the packages it contains — so, for example, the base seed will contain a minimum set of packages needed to get a minimal command shell on a device.

The seeds expand into a list of packages that are then built using some automated scripts to create CD and DVD images. These images are then published daily on cdimage.ubuntu.com and eventually as releases on releases.ubuntu.com. The types of images that are created are

- **.iso images** — Ubiquity (desktop) and the alternate installer
- **.img images** — Used for ARM and for LPIA images.

Inside Seed Germination

Germinate is a program that constructs lists of packages by reading seed files. Seed files are files that contain raw lists of packages. These raw lists are processed by the germinate program to generate another list, which contains all of the dependencies of the packages in the raw list (Germinate also creates separate lists of *suggests* and *recommends* in its output). An example germinate output illustrated in Figure 11-1 shows this — which is from the "minimal" seed output.

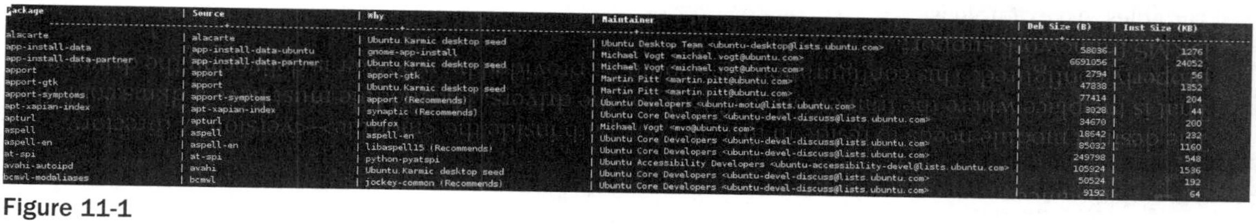

Package	Source	Why	Maintainer	Deb Size (B)	Inst Size (KB)
alacarte	alacarte	Ubuntu.Karmic desktop seed	Ubuntu Desktop Team <ubuntu-desktop@lists.ubuntu.com>	58036	1276
app-install-data	app-install-data-ubuntu	gnome-app-install	Michael Vogt <michael.vogt@ubuntu.com>	6691056	24052
app-install-data-partner	app-install-data-partner	Ubuntu.Karmic desktop seed	Michael Vogt <michael.vogt@ubuntu.com>	2794	56
apport	apport	apport-gtk	Martin Pitt <martin.pitt@ubuntu.com>	47838	1352
apport-gtk	apport	Ubuntu.Karmic desktop seed	Martin Pitt <martin.pitt@ubuntu.com>	77414	204
apport-symptoms	apport-symptoms	apport (Recommends)	Ubuntu Developers <ubuntu-motu@lists.ubuntu.com>	3028	44
apt-xapian-index	apt-xapian-index	synaptic (Recommends)	Ubuntu Core Developers <ubuntu-devel-discuss@lists.ubuntu.com>	34670	200
apturl	apturl	ubufox	Michael Vogt <mvo@ubuntu.com>	18642	232
aspell	aspell	aspell-en	Ubuntu Core Developers <ubuntu-devel-discuss@lists.ubuntu.com>	85032	1160
aspell-en	aspell-en	libaspell15 (Recommends)	Ubuntu Core Developers <ubuntu-devel-discuss@lists.ubuntu.com>	249798	548
at-spi	at-spi	python-pyatspi	Ubuntu Accessibility Developers <ubuntu-accessibility-devel@lists.ubuntu.com>	105924	1536
avahi-autoipd	avahi	Ubuntu.Karmic desktop seed	Ubuntu Core Developers <ubuntu-devel-discuss@lists.ubuntu.com>	50524	192
bcmwl-modaliases	bcmwl	jockey-common (Recommends)	Ubuntu Core Developers <ubuntu-devel-discuss@lists.ubuntu.com>	9192	64

Figure 11-1

In the preceding table, aptitude is included because tasksel depends on it, and busybox-initramfs is included because initramfs-tools depends on it.

The raw seed file sometimes begins with headers at the top of the file, in "key: value" format. These are not parsed by germinate but are used by the program germinate-update-metapackage when generating its output. The packages in the raw seed lists should be listed in bullet form with each package preceded by an asterisk. The raw seed file itself can be written in wiki format (as seen below) because everything that does not start with an asterisk will be ignored.

```
== Development ==

These packages are needed in order to build Ubuntu packages.

 * build-essential
 * fakeroot

== Hardware & Network Access ==

 * pptp-linux           # client for Microsoft-compatible VPN's, needed for some ISPs
 * sl-modem-daemon      # needed for some Winmodems (see OutoftheboxWinmodem)
 * bpalogin
 * ndiswrapper-utils-1.9 [amd64 i386]
 * ndisgtk
```

In the preceding example, pptp-linux is a dependency. If it were in brackets, it would indicate that the package is recommended and not depends. For instance, the Online Services example that follows is a recommends:

```
Online Services:
 * (ubuntuone-client-gnome)
```

A leading exclamation mark in a seed file indicates a per-seed blacklist. An exclamation mark means to never include this package in the seed or in any of its inner seeds. An example of this follows:

```
== Blacklist ==
libavcodec cannot be shipped on CDs (c.f. Ubuntu technical board resolution
2007-01-02).
!libavcodec*
```

The asterisk at the end of libavcodec is used for regular expression matching in the germinate code to place everything that starts with libavcodec into the per seed blacklist.

Other things to notice in seed files are process substitution variables. A package containing substitution variables will be expanded into one package for each possible combination of the values of those variables. An example is:

```
 * Kernel-Stem: linux linux-image

== i386 ==

 * ${Kernel-Stem}-generic-pae [i386]
 * linux-headers-generic-pae [i386]
 * ${Kernel-Stem}-virtual [i386]
 * linux-headers-virtual [i386]
```

Seed entries can also be followed by square brackets. An example is:

```
 * language-support-${Languages} [!powerpc]
```

which indicates that the language-support package should not be used on the powerpc architecture. The converse is also true, of course, in that seed entries followed by `[lpia i386]` (without the exclamation mark) indicate that they should only be used on LPIA and i386 architectures.

Finally, if the package name in the seed starts with % then germinate expands to all binaries from the given source package.

Germinating the Seeds

As mentioned previously, the raw seed files are germinated into a full list of packages using the germinate program. The germinate program requires some default seed files to be present in order to work. The most important of these seed files (which are mandatory) are STRUCTURE, supported, and blacklist.

These mandatory files do not necessarily need to contain anything but they must be present for germinate to work.

Chapter 11: Putting It All Together

STRUCTURE is a special file that lists the seeds and how they depend on each other. The STRUCTURE file for karmic netbook looks like this:

```
include platform.karmic
netbook-remix: desktop-common
live: netbook-remix
ship-live: live
supported: netbook-remix ship-live supported-common
```

The name of the seed file is followed by a colon. After the colon is a space and a space-separated list of seeds on which the first seed on the line depends. It is possible to see graphically how the seeds relate to each other by looking at the `structure.dot` file in the germinate output. Figure 11-2 shows the graphical germinate output for Ubuntu Netbook Remix, which is available from `http://people.ubuntu.com/~ubuntu-archive/germinate-output/unr.karmic/structure.dot`.

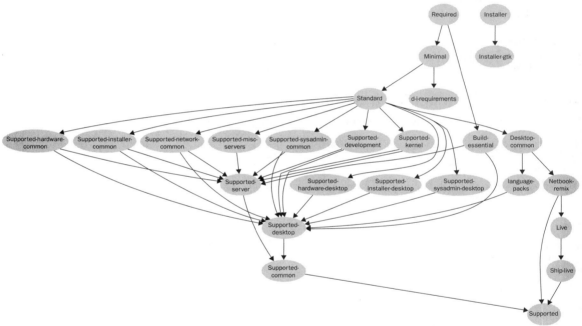

Figure 11-2

A STRUCTURE file then lists the inheritance relationships between seeds. It is an ordered sequence of lines like the following:

```
SEED:[INHERITED]
```

Chapter 11: Putting It All Together

INHERITED is a space-separated list of seeds from which SEED inherits. For example, `netbook-remix: desktop-common` in the preceding STRUCTURE file indicates that packages in the `netbook-remix` seed may depend on packages in the `desktop-common` seed without requiring that those packages appear in the netbook-remix output. Any line like the following:

```
include BRANCH
```

which is seen in the section below, causes another seed branch to be included.

The *supported* seed lists all the packages that are necessary to build the distribution (but which do not necessarily need to be included on the final image for space reasons). germinate, therefore, does not add all the packages that result from following build-dependencies of seed packages and of their dependencies to every output, unless they are also in the seed list. These packages are added to supported. Like any other seed, the supported seed may contain its own list of packages.

The *blacklist* file does not define a list of packages to include but lists packages that will never be included in the output of germinate.

The other seeds that were listed in the STRUCTURE file in the Ubuntu Netbook Remix example, such as netbook-remix, desktop-common, live, ship-live, and so on, can be called anything but they need to exist, and they need to be listed as such in the STRUCTURE file.

An Example: Germinating Ubuntu Netbook Remix

Begin by branching the Ubuntu Netbook Remix seed. Make a seeds directory:

```
$ mkdir seeds; cd seeds
$ bzr branch lp:~ubuntu-core-dev/ubuntu-seeds/unr.karmic
```

move into the unr.remix folder, and in the STRUCTURE file there is a line like the following:

```
include platform.karmic
```

which means that the platform karmic seed is required. Move back to the seeds folder and branch platform.karmic:

```
$ cd ..
$ bzr branch lp:~ubuntu-core-dev/ubuntu-seeds/platform.karmic
```

Next make an output folder and move into it the following:

```
$ mkdir output; cd output
```

Make sure that the germinate tool is installed (sudo apt-get install germinate) and then invoke it:

```
$ germinate -S file:///home/<user>/Dev/Book/test/seeds/ -s unr.karmic -m
http://172.16.50.131:9999/ubuntu -d karmic -a i386 -c main,restricted
,universe,multiverse
```

172.16.50.131:9999 is our local mirror — use archive.ubuntu.com if you do not have a local mirror set up.

Chapter 11: Putting It All Together

Using seeds and germinate creates a final package list with all the other packages that are required, including all of their dependencies.

Packages and Repositories

This section deals with creating meta packages and creating a local repository for your content.

Generating Metapackages the Ubuntu Way

To create a metapackage the Ubuntu way, get the source of the ubuntu-netbook-remix for Ubuntu Karmic:

```
$ apt-get source unr-meta
```

Do not forget to add karmic deb-src repository to your sources.list.

Make sure that the germinate tool is installed:

```
$ sudo apt-get install germinate
```

Move into the unr-meta directory, which was created when we checked out the source of unr-meta, and remove metapackage-map and all the netbook-remix-* files

```
$ rm metapackage-map netbook-remix-*
```

Edit the update script and remove the `--bzr` parameter from the last line:

After the changes, the last line looks like this:

```
$ exec germinate-update-metapackage
```

This is important only when the seeds are on a local machine and not in a Bazaar repository. If the seeds are in Bazaar this parameter can be added again.

The --bzr parameter forces the germinate-update-metapackage to download the seeds files from a Bazaar repository. If this is not given, it will fall back to the ubuntu seed repository.

Next, edit the file update.cfg. This is the configuration file for germinate-update-metapackage and, by default, the file looks like this:

```
[DEFAULT]
dist: karmic

[karmic]
seeds: netbook-remix
architectures: i386 amd64 powerpc ia64 sparc lpia armel
seed_base: http://people.ubuntu.com/~ubuntu-archive/seeds/
archive_base/default: http://archive.ubuntu.com/ubuntu/
```

235

Chapter 11: Putting It All Together

```
archive_base/ports: http://ports.ubuntu.com/ubuntu-ports/
archive_base/hppa: %(archive_base/ports)s
archive_base/ia64: %(archive_base/ports)s
archive_base/lpia: %(archive_base/ports)s
archive_base/powerpc: %(archive_base/ports)s
archive_base/sparc: %(archive_base/ports)s
archive_base/armel: %(archive_base/ports)s
components: main restricted

[karmic/bzr]
seed_base: bzr+ssh://bazaar.launchpad.net/~ubuntu-core-dev/ubuntu-seeds/
seed_dist: unr.%(dist)s
```

After the changes, the file should look like this (read the comments):

```
[DEFAULT]
dist: karmic

[karmic]
seeds: netbook-test
architectures: i386
# put here the path to your seed files. In this path there should be platform
# .karmic and netbook.karmic.
seed_base: file:///home/user/devel/path/to/seeds/
# Here we have the repositories separated by commas. The first repository is a
# caching
# debian repository and the second is a local repository that will contain the
# modified packages.
archive_base/default: http://localhost:9999/ubuntu/, file:/home/user/devel/path/to/
repos
# Here will be set the name of the dir that contains the seed files
seed_dist: netbook.%(dist)s
```

Then run the update script:

```
$ ./update
```

This script calls germinate-update-metapackage, which constructs the metapackage.

germinate-update-metapackage renames the current directory, updates the debian/control to reflect the current contents of the seeds, and updates debian/changelog with a description of the changes it made. The package can now be manually rebuilt using dpkg.

Building the metapackage

Rename the unr-meta directory to a new name. This will be the new name for the source package. An example might be renaming unr-meta-1.157 to netbook-meta-1.157.

This is renaming the source *package. It is not the name of the new metapackage. This is named in the second stanza of the debian control file.*

Chapter 11: Putting It All Together

Next, move into the netbook-meta-1.157 directory and then into the debian subdirectory. Edit the control file and replace the old package name (for example, ubuntu-netbook-remix) with the new metapackage name (for example, ubuntu-netbook-example). An example control file looks like the following:

```
Source: netbook-meta
Section: metapackages
Priority: optional
Maintainer: Ubuntu Mobile Developers <ubuntu-mobile@lists.ubuntu.com>
Standards-Version: 3.7.3
Build-Depends: debhelper (>= 4), germinate (>= 0.42)

Package: ubuntu-netbook-example
Architecture: any
Depends: ${ubuntu-netbook-example:Depends}
Recommends: ${ubuntu-netbook-example:Recommends}
Description: The Ubuntu Netbook Example system
 This package depends on all of the packages in the Ubuntu Netbook Remix
 System
 .
 It is also used to help ensure proper upgrades, so it is recommended that
 it not be removed.
```

Save this file and then edit the debian/rules file. If you changed the main seed filename in the STRUCTURE file in the section "Germinating the Seeds" (for example, from netbook-remix to netbook-example), you should replace this, too. Here is an example rules file:

```
#!/usr/bin/make -f

clean:
dh_testdir
dh_clean
rm -rf build-stamp *.old debootstrap-dir

DEB_BUILD_ARCH ?= $(shell dpkg-architecture -qDEB_BUILD_ARCH)

build: build-stamp
build-stamp: netbook-example-$(DEB_BUILD_ARCH)
dh_clean
for seed in netbook-example; do \
package=ubuntu-$$seed; \
(printf "$$package:Depends="; perl -pe 's/\n/, /g' $$seed-$(DEB_BUILD_ARCH); echo) \
>> debian/$$package.substvars; \
(printf "$$package:Recommends="; perl -pe 's/\n/, /g' $$seed-recommends-$(DEB_BUILD_ARCH); echo) \
>> debian/$$package.substvars; \
Done
touch $@

install: build-stamp

binary-arch: install
dh_testdir -a
```

Chapter 11: Putting It All Together

```
    dh_testroot -a
    dh_installdocs -a
    dh_installchangelogs -a
    dh_compress -a
    dh_fixperms -a
    dh_installdeb -a
    dh_gencontrol -a
    dh_md5sums -a
    dh_builddeb -a

binary-indep:
    dh_testdir -i
    dh_testroot -i
    dh_installdocs -i
    dh_installchangelogs -i
    dh_compress -i
    dh_fixperms -i
    dh_installdeb -i
    dh_gencontrol -i
    dh_md5sums -i
    dh_builddeb -i

binary: binary-indep binary-arch

.PHONY: binary binary-arch binary-indep clean checkroot build
```

Finally, describe the changes made to debian/changelog using the dch tool. From the root directory of the metapackage type the following:

```
$ dch -i
```

This will open the debian/changelog using your default editor. Update the debian/changelog with a description of the changes made. An example changelog looks like this:

```
mobile-example (1.104) karmic; urgency=low
  * Fork ubuntu-meta source as mobile-example.
  * Cleanup rules.
  * Set Maintainer to Mobile Developer <developer@mobile.org>.
  * Switch seed_base to ~ubuntu-mobile/ubuntu-seeds and seed_dist to
    mobile.%(dist)s
```

The package can now be manually rebuilt using `debuild`.

Generating Metapackages the Simple Way

It is possible to run the following command on a device whose installed software packages you wish to copy. This type of thing is known in Brazil as a gambiarra.

```
$ dpkg-query -W -f='${Package}, ' | sed 's/\, $//' > metapackage-list.txt
```

Chapter 11: Putting It All Together

Here is a snippet from running this command on an Acer Aspire device running karmic:

```
acpi-support, acpid, adduser, alacarte, alsa-base, alsa-utils, anacron, apmd, app-
install-data, app-install-data-commercial, apparmor, apparmor-utils, apport,
apport-gtk, apt, apt-transport-https, apt-utils, apt-xapian-index, aptitude,
apturl, aspell, aspell-en, at, at-spi, avahi-autoipd, avahi-daemon, avahi-utils,
base-files, base-passwd, bash, bash-completion, bc, bind9-host, binfmt-support,
binutils, bluetooth, bluez, bluez-alsa, bluez-cups, bluez-gnome, bluez-gstreamer,
bluez-utils, bogofilter, bogofilter-bdb, bogofilter-common, brltty, brltty-x11,
bsdmainutils,
```

The preceding command will query the dpkg database and return a comma-delimited list of all packages that can be added into the `Depends` portion of a control file.

Once the control file is written (a metapackage just needs to have a control file), the metapackage can be built using the following:

```
$ dpkg-deb -b meta-package-name.deb
```

Preseeding the Installer

This sections explains what it means to preseed an installer.

As explained previously, two installer types are available on Ubuntu — the first is the debian-installer and the second is Ubiquity. With the debian-installer, a full-featured preseed is available, whereas with Ubiquity there are some restrictions on what can be done with preseeding.

If no preconfiguration is necessary, all that is needed is to change the name of the main package, so, for example, changing the following preseed file found at ubuntu-cdimage/debian-cd/data/karmic/preseed/ubuntu-netbook-remix/netbook-remix.seed

```
tasksel tasksel/first multiselect ubuntu-netbook-remix
d-i     preseed/early_command    string . /usr/share/debconf/confmodule; db_get
debconf/priority; case $RET in low|medium) db_fset tasksel/first seen false ;; esac
d-i     passwd/auto-login        boolean true
```

to the new file found at ubuntu-cdimage/debian-cd/data/karmic/preseed/ubuntu-netbook-example/netbook-example.seed, looks like this:

```
tasksel tasksel/first multiselect ubuntu-netbook-example
d-i     preseed/early_command    string . /usr/share/debconf/confmodule; db_get
debconf/priority; case $RET in low|medium) db_fset tasksel/first seen false ;; esac
d-i     passwd/auto-login        boolean true
```

If preconfiguration is necessary, then options can be added into this file. An example preconfiguration file that you can use as the basis for your own is available from http://d-i.alioth.debian.org/manual/example-preseed.txt.

Various options are available. Examples include setting a default locale for the installation

```
d-i debian-installer/locale string pt_BR
```

Chapter 11: Putting It All Together

installing non-free firmware

```
d-i hw-detect/load_firmware boolean true
```

or even reporting back the most popular software on a device:

```
popularity-contest popularity-contest/participate boolean true
```

It is also possible to create your own example preseed file by using debconf-get-selections. This will output a list of all debconf options you've chosen throughout the install on a device.

```
$ sudo apt-get install debconf-utils
$ sudo debconf-get-selections --installer > netbook-example.seed
$ sudo debconf-get-selections >> netbook-example.seed
```

Adding Packages to the Image

Copy the packages you wish to add to the image to

```
LOCALDEBS=$CDIMAGE_ROOT/local/packages
```

and then add to your build image command below the parameters:

```
LOCAL=1 LOCALDEBS=$CDIMAGE_ROOT/local/packages
```

Finally, Build the Custom ISO

With the new seed list, everything is ready to build the final custom image:

```
$ LOCAL=1 LOCALDEBS=$CDIMAGE_ROOT/local/packages PROJECT=ubuntu CAPPROJECT=Ubuntu ↵
DIST=karmic ARCHES=i386 CDIMAGE_NOSYNC=1 IMAGE_TYPE=daily build-image-set daily
```

This can then be distributed and installed on the target device.

Ubuntu Policies, Trademarks, Copyright, and Common Sense

It is important to note that Ubuntu mirrors are open and accessible to everyone who wishes to download the data they hold. As such, they are a public service for anyone to use. Having said this, however, there are some things that should be done out of courtesy when creating a derived distribution.

What follows should be considered only an outline of what is expected when creating a derivative distribution. It is always a good idea to consult with Canonical Ltd. before releasing anything into the wild that you are unsure about. The following things are important:

Chapter 11: Putting It All Together

- You should change the DISTRIB_RELEASE variable in the file /etc/lsb-release to something compatible with the derived distributions naming scheme.

- You should change the Ubuntu logo in Ubiquity. This could also include changing the new Ubiquity slideshow if this is included in the distribution (bzr branch lp:ubiquity-slideshow-ubuntu).

- You should state publicly that the distribution is based on Ubuntu.

- You should Give feedback to Ubuntu. Provide bug reports, although patches and documentation are even better!

So What Is a Derived Distribution?

Confusion exists over the exact definition of the term "derivative" or *derived* distribution. Deriving some new software from some original implies that some sort of change has taken place in the original — if this were not the case, then it would just be a simple gcc compile of the original source code (something called a *compilation* in copyright law). Obviously, if the original source code is changed, then the upstream author has the right to insist that the new work be licensed under the same provisions as the original code. This is clear — however the confusion comes with the definition of *a new work*.

Copyright law also allows for a situation where if the original source code is "annotated," "elaborated," or "modified" it can be considered, *on the whole* a new work. This new work is then subject to the same license terms as the original source code mentioned earlier.

The problem is with the phrase "on the phrase" — "on the whole," in English, is used most frequently to mean "generally" which seems awfully vague for a legal clause in a copyright contract. As a consequence of this, a new work *might* be seen as a derived work if creative expression is added to the original work, even if there are no actual changes to the original source code.

Some people think that this doubt has led to more interest in using other licenses (such as the LGPL), which explicitly allow for derivative works rather than relying on the GPL.

When to Use the LGPL

The original name for this library was " The Library General Public Licence, " because this license covers the linking of libraries between applications. Today it is more commonly referred to as " The Lesser General Public License. "

Imagine that there are two applications. When both applications are licensed under the GPL, the source code must be made available — full stop. However, the situation is complicated when one is GPL and the other a proprietary application, and even more so when the proprietary application makes use of some library from the GPL one. The LGPL license specifically allows for the linking of a proprietary application to GPL ' d libraries without requiring that the proprietary application make its source code available. If the proprietary application changes some of the source code in the GPL ' d library (perhaps to make the application bind better), then only *those changes* need to be published. In practice and as a matter of courtesy, companies should do their utmost to make sure that this code is accepted upstream as quickly as possible. It is far easier to maintain this way, ensures future application compatibility, and makes community relationships stronger.

Chapter 11: Putting It All Together

Some people believe the fact that proprietary applications (as well as GPL'd ones, of course) can be derived from LGPL'd code increases the developer base and business ecosystem around the library. This strategy is being followed by both Intel with its Clutter library (see Chapter 4, "Application Development") and Nokia, who recently purchased Trolltech (the makers of QT) and then LPGL'd the whole QT library.

Other people are not so convinced about the community building benefits, hence the common name of "Lesser." Once the copyright has been correctly assigned, it is possible to go on and actually build the package.

> *The ISO image (the distribution) we made in the section "Building a Customized Ubuntu Image" is copyrighted by the authors and not GPL. The reason for this is that this distribution constitutes an aggregated or collective work, with all the individual applications included in our custom distribution licensed under their respective licenses (proprietary, GPL, or whatever) and the actual seed selections themselves under our copyright.*

Summary

In this chapter, we built a custom Ubuntu image that is based on Ubuntu Karmic.

It was made possible through the work of countless Debian and Ubuntu developers, community members, and technical writers. The fact that this can be done at all is a tribute to the distributed nature of open source development and a credit to everyone everywhere who has in some way contributed to free software development.

12
Mobile Directions

> *Architecture: A fundamental underlying design of computer hardware, software, or both.*
>
> www.dictionary.com

The marketplace has opened up a great number of developer opportunities across the mobile landscape. Linux and Ubuntu are in a great position to be widely used in mobile applications. The precise confluence of hardware, software, and mobile/network technologies is unpredictable, so it pays to be prepared with the skills and capabilities to move in various directions as future scenarios unfold. Working through this book is a great part of that preparation. Cutting through some of the market chaos could also help.

As you plan your own mobile activities, a roadmap may prove useful. The architectural block diagram shown in Figure 12-1 offers one possibility of what a mobile computing platform might look like.

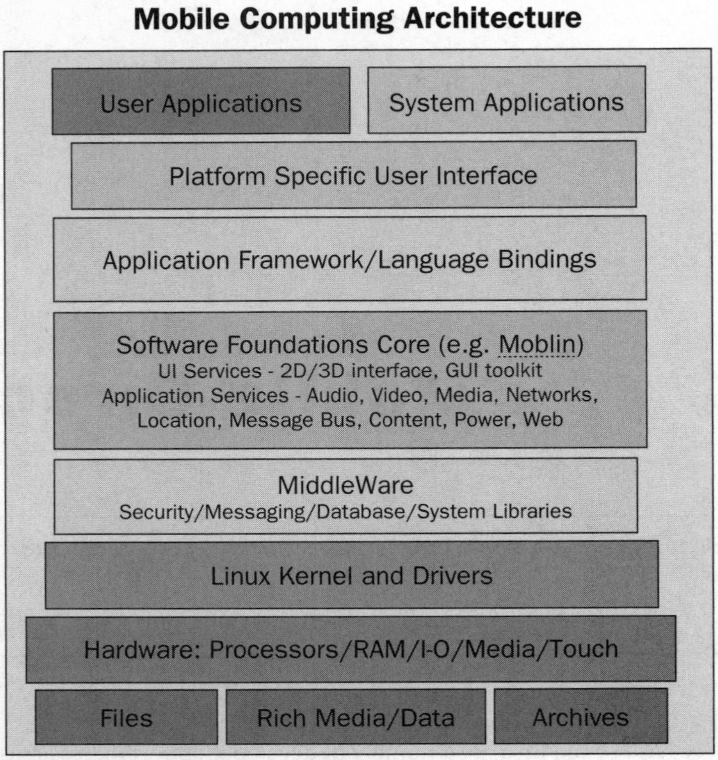

Chapter 12: Mobile Directions

Mobile Computing Architecture

- User Applications
- System Applications
- Platform Specific User Interface
- Application Framework/Language Bindings
- Software Foundations Core (e.g. Moblin)
  UI Services - 2D/3D interface, GUI toolkit
  Application Services - Audio, Video, Media, Networks,
  Location, Message Bus, Content, Power, Web
- MiddleWare
  Security/Messaging/Database/System Libraries
- Linux Kernel and Drivers
- Hardware: Processors/RAM/I-O/Media/Touch
- Files | Rich Media/Data | Archives

Figure 12-1

The subsystems are moving targets — and under the control of far-flung, independent individuals, groups, and foundations. Development is not always guided by market incentives, nor marked by a corporate calendar of release cycles. The existing components need to be maintained, and new components will be invented. New hardware will keep driving change through the software stack. Virtually all of open source is built with a common mission of sharing and improvement. As a member of the open source community, you can directly impact virtually any one of these key subsystems.

Choice, Change, and Opportunity

Norbert Wiener was a world-renowned mathematician. A child prodigy, he graduated from college at age 14 and earned a PhD from Harvard at 18. He spent most of his career teaching at MIT and so he knew the halls of that institution pretty well. He wrote *Invention: The Care and Feeding of Ideas* in the 1950s. Wiener described four key stages of invention: intellectual climate, technical climate, social climate, and economic climate.

Wiener emphasized the need to encourage and recognize invention and the creation of ideas. He also recognized the burden and responsibility of carrying these inventions forward in trusting stewardship

Chapter 12: Mobile Directions

for a community. In many ways, Wiener's writing foresaw the open source revolution and the role of Linux distributions.

By Wiener's account, the risks are great in the early inventive stages and simply cannot be calculated. The inventive mind can do nothing more than follow instinct, work with knowledge, and act.

Moving beyond that difficult first step, Wiener doesn't hide the fact that really big developments often take a long time to mature. All of us have witnessed enormous technological changes that the Internet, the World Wide Web, and computers have brought during the last three decades.

When contemplating your next steps — choice, change, and opportunity — think of these words from Weiner for inspiration around Ubuntu, Linux, Mobile computing, and open source:

> *The profit motive may be important, but it must be supplemented by other motives. The community must cultivate a group which is neither subservient to the profit motive in the external community nor internally governed by this motive. . . . whatever benefits are awarded for scientific creation should have the good of the community as their purpose even more than the good of the individual. As such, they should be contingent on a full and free publication of the new ideas of the discoverer. The truth can make us free only when it is a freely obtainable truth.*

The Ubuntu Developer Community that appears in Figures 12-2 and 12-3, gathers at different locations around the world to plan out each new release. Keep an eye on the schedule for the next Ubuntu Developer Summit near you and participate!

Figure 12-2

Chapter 12: Mobile Directions

Figure 12-3

Evolution and Software Development

Linus Torvalds defines this kind of inventive and creative spirit. If you want an excellent read and a detailed look at how invention works, check out *Just for Fun: The Story of an Accidental Revolutionary* by Linus Torvalds and David Diamond. The book begins in the early days, before Linus became Linux, and it's a page turner. By the time you're done, you'll have one of the best narratives on how technology development really happens. The hardware equivalent of a story like this would be Tracy Kidder's Pulitzer Prize winning "Soul of a New Machine."

Torvalds' book is fun to read, but it's also a learning lab look at how great open source technologies evolve. In the case of Linux, the project started out small, with few expectations. It quickly grew beyond something that the imagination (and even big corporations) could comprehend.

There's a good chance that somewhere in the world, another young software genius or engineer is working on something that will have at least as much an impact on the computer industry in the next decade or two.

Chapter 12: Mobile Directions

While Torvalds was working on his first lines of Linux code, Richard Stallman was marching on Lotus Development and protesting for Free Software. That was around 1990, when I worked for Lotus Development. I looked out my office window one day and saw Stallman and others in front of Lotus's main office building. I distinctly remember one protestor who held a sign with something improbable written on it: To paraphrase, it said "All Software Should Be Free."

At the time, it was difficult to comprehend how any rational individual could expect that complex and high quality software such as Lotus 1-2-3 could be developed for free. (Of the more than 100 people working on Lotus 1-2-3, at least some had to pay room and board. Richard Stallman, on the other hand, describing himself as a "squatter," seemed to manage a rent-free office and free overnight sleeping accommodation at MIT, in addition to commanding his $260/hour consulting fee!)

Over the next ten years, a great deal of tiny, gradual change swept through the tech industry. All around the world, business models and environments changed — usually gradually but then explosively in the fall of 2008. Technology today is in a very different place than it was ten years ago, and even more evolved from where it was 20 years ago. As a method for explaining how some of this evolution might be discussed, I offer four evolutionary software development models for your consideration. All of the models are inspired by various works of natural philosophers. I call these models Darwin, Mendel, Mayr, and Frankenstein.

Darwin

The *Darwin Model of Software Development* is named for Charles Robert Darwin, famous for his publication of "On the Origin of Species." Darwin published his breakthrough book at the wise age of 51. Darwin's book documents his lifetime of research and formulation of theories on "Natural Selection." Wikipedia provides a nice concise description of this research and discoveries: "Natural Selection is the process by which favorable heritable traits become more common in successive generations of a population of reproducing organisms, and unfavorable heritable traits become less common, due to differential reproduction."

Applying the Darwin Model helps explain why more people use Windows than DOS, why Linux has become more prevalent than UNIX, and why notebooks replaced laptops (which previously supplanted luggables). The Darwin Model explains why computers don't need to be 50 feet long and 8 feet tall. As DeMoivre wrote in his seminal statistics book, "The Probability of an Event is greater or less, according to the number of Chances by which it may happen, compared with the whole number of Chances by which it may either happen or fail." Because there has been a recent rush of mobile technologies — hardware and software — and open source offers virtually infinite combinations, there is a high probability some wonderful mobile developments will "happen."

Mendel

The *Mendel Model* of software development is named after the person considered the Father of Modern Genetics, Gregor Johann Mendel. In a paper written by Mendel, "Experiments on Plant Hybridization," people caught the first glimpse of how dominant and recessive genes worked. Mendel was a clever part-time scientist and learned how to create "best of breed" outcomes. Wikipedia describes Mendelian inheritance as "a set of primary tenets relating to the transmission of hereditary characteristics from parent organisms to their children."

Chapter 12: Mobile Directions

Linux-based distributions are careful, selective, scientific, and personalized works of many people individually "breeding" their own distributions. Creating a distribution is essentially a work of Mendelian selection. The individual programs, packages, and customizations in a distro are tailored to the preferences and needs of the people who are creating the distribution.

Mendel could certainly trace back the inherited traits of Ubuntu from Debian. Mark Shuttleworth got involved with the Internet as a student. He became a Debian developer by 1994, after he created the first package for Apache. From that point forward, Debian became an anchor and foundation for his work. Shuttleworth was the first "Master of the Universe" for Ubuntu — the group of individuals who edit, maintain, include, exclude, and identify interdependencies in the components of a new Ubuntu release.

Mayr

The *Mayr Model* of software development is named after Ernst Mayr. Mayr came to be known as the leading evolutionary biologist. His great work revealed the intricacies of adaptation and multiplication of species. Mayr's work harmonized and connected the work of Darwin and Mendel, bringing together previously warring camps of scientists. Mayr's theories built bridges rather than walls. Mayr was well known for the way in which he recruited and mentored other interested scientists through bird watching societies.

The Mayr Model explains the co-existence of diverse Linux distributions. In many ways, Mayr's methodology mirrored that of open source development. Through collaboration and knowledge sharing, Mayr brought together individuals from very different areas with common interests. There is another similarity to Mayr worth noting. Eric Raymond memorably wrote that "Every good work of software starts by scratching a developer's personal itch." Echoing that refrain, Mayr was known to encourage his students to take up their own personal research projects: "Everyone should have a problem."

Frankenstein

Finally, there's the *Frankenstein Model* of software development. As you might guess, this is based on the work of Mary Shelly's *Frankenstein or The Modern Prometheus*, which is her 1818 novel that was inspired in part by discussions she had about the work of Erasmus Darwin, grandfather of Charles Darwin. The plot of the story is simple — a scientist named Frankenstein discovers he has the ability to animate life during his college days. (A typical college awakening perhaps after too many nips at the pub?) Frankenstein then brings to life, from a patchwork of body parts, a new being. Does this quote from the book sound familiar to any open source developers?

> *I had worked hard for nearly two years, for the sole purpose of infusing life into an inanimate body. For this I had deprived myself of rest and health. I had desired it with an ardor that far exceeded moderation; but now that I had finished, the beauty of the dream vanished, and breathless horror and disgust filed my heart.*

The Frankenstein Model is descriptive of how outcomes sometimes — despite the best of intentions — come up short of original expectations. These are the projects that don't end well for anyone. Yet, the code still lives on. If there is a lesson to be learned, Frankenstein the scientist puts it this way: "Peace! Peace! Learn my miseries, and do not seek to increase your own."

There is no fooling the future; it is what it will become. However, these different models might help you understand the past, and make some well-informed guesses about the future.

Big Ideas to Think About

All of us live in a dynamic time that is influenced by global cultures, politics, economics, and technologies. The only sure thing is that we must expect the unexpected. Here are some big ideas to think about, where you can be sure to expect some unexpected influences on mobile computing, Linux, and Ubuntu.

The Politics of Technology

President Barack Obama leveraged the Internet during his campaign in ways that certainly helped him win the election. (He has more than 4.5 million friends on Facebook!) During his first month in office, the new administration requested a report on the use of open source in government. As the New York Times described it, he "has been all but addicted to his BlackBerry." Obama is the most tech savvy president the United States has ever had, and that is bound to influence policy.

At the same time, across the pond, the UK educational system is discussing open source. From the BBC on January 26, 2009:

> With Open Source Software (OSS) freely available, covering almost every requirement in the national curriculum, a question has to be asked why schools do not back it more fully, possibly saving millions of pounds.
>
> As the name suggests, OSS is community-driven software with its source code open to all. Anyone can modify the software according to their needs and then share these modifications with everyone else.

How will the politics of technology shape our mobile future? What impact will government decisions and policies have on the production of open source and the use of Ubuntu?

The Next Billion

MIT's Media Lab has an initiative known as The Next Billion Network. It presents a rather startling scenario related to mobile computing:

> Within the next three years, another billion people will begin to make regular use of cell phones, continuing the fastest adoption of a new technology in history. Soon, this next billion will make its voice heard — and connect to the global information network.
>
> http://nextbillion.mit.edu/

With certainty, this will have an impact on Ubuntu, Linux, and mobile computing.

Sensory Overload

GPS information, WiFi/cell phone positioning, and other means of geographically aware technology can certainly help make mobile computers more location aware. Will new technologies be developed to fill in the gaps of GPS and WiFi/cell phone positioning? What privacy issues will we face?

Chapter 12: Mobile Directions

Cloud Computing

We are seeing cloud computing reach out to consumers and corporations. People now routinely use the Cloud to store e-mail as webmail. Google has a full range of office applications available online. Twitter, FriendFeed, Facebook, and other social network tools maintain sensitive personal data on the Web. Cloud computing can eliminate or reduce the step of data synchronization and massive file copying between mobile and fixed location computer systems. Will this drive faster adoption to handheld mobile devices such as MIDs? With so much private data on the Web, are there new security measures and standards that must be considered and incorporated into software stacks?

ARM Wrestling

It looks like a battlefield is forming along the lines between x86 and ARM architectures. The puns seem endless, like "ARM Wrestling" or "A shot in the ARM for Intel." In the long-run, this should be good for the consumer/user, and prove healthy for the industry. Competition creates innovation and accelerates improvement. However, the threat to Intel's x86 processor dominance is quite real. From *The New York Times* on June 30, 2008:

> In addition to Qualcomm and Nvidia, there are more than 200 licensees of the ARM processor design, including major chip makers like Marvell and Texas Instruments. Together, they supply the more than 1.1 billion cellphones, many of which use multiple ARM chips. The chips are also used in a growing array of special purpose consumer electronics like G.P.S. navigators and set-top TV boxes.

There are a lot of ARMs out there! Here's another data point, from the ARM Wikipedia entry:

> As of January 2008, over 10 billion ARM cores have been built, and iSuppli predicts that 5 billion a year will ship in 2011.

I've installed Ubuntu on both ARM-based systems and Atom-powered systems. I like the processing power of the Atom, but I like the low-power consumption of the ARM. This will be an exciting battle to participate in.

At the same time, an important new technology foundation to mobile solutions is being developed. The Moblin initiative, for example, has growing support from hardware and software companies after it received initial funding from Intel. In April 2007, Intel first introduced the concept of a "MID" computer at its Developer Conference in China. In July 2007, Intel formalized a software development initiative under the name of Moblin and launched a website at Moblin.org. (The name Moblin is a contraction of mobile and Linux.)

In October 2007, I attended the Ubuntu Developer conference for several days of sessions on mobile computing. Canonical staff and Intel participants mainly led those sessions, with a number of other open source contributors also in attendance. After an initial release of Moblin V1 based on Ubuntu, Intel directed further explorations to Fedora distributions. That provided another fruitful source of innovation and code. Intel pressed on and late in 2008 announced plans for the Moblin V2 release. Just about a year later, October 30, 2008, Intel issued a press release on its plans to support and accelerate growth of mobile device makers in Asia. Intel announced joint plans with Taiwan's Ministry of Economic Affairs to create a Moblin laboratory for development, training, and support of hardware manufacturers.

Chapter 12: Mobile Directions

As Ars Technica wrote at the time:

> *Intel is putting a lot of resources into advancing mobile Linux adoption and making Moblin a solid platform for device makers. If Moblin can resolve some of the problems that have afflicted Linux device makers and offer a more cohesive platform specifically for Atom-based devices, it could be a big win for the open source operating system.*

In January 2009, Intel released the first Alpha of its Moblin V2 Core release targeted at netbooks. A follow-up release appeared in March. One of the widely commented on features is "Fastboot," which gets netbooks up and running fast!

In the future, there will be multiple processor platforms (e.g., ARM, x86) and several distribution pathways (e.g., Ubuntu, Moblin, and others) competing for developer attention.

Razors and Blades

What would happen if microprocessor chips and their packaging fell to a price point where it was practical for them to be modular and disposable? If computer components were modular enough, they might be routinely swapped, exchanged, or updated.

Rather than selling entirely new computers (a transaction that is neither cheap nor environmentally friendly or time efficient), companies might consider business models around manufacturing that could support modular "razors and blades" markets. For instance, a company might sell you an attractively designed razor at a subsidized discount, knowing you'd be locked into buying its brand of blades. Computer printers and inks may follow a similar business model. Why not microprocessors?

Free Lunch

One of the original tenets of open source software, according to gnu.org, was this: "Free software is a matter of liberty, not price. To understand the concept, you should think of free as in free speech, not as in free beer."

In the early days of open source, "free speech" was an easier argument to make in the face of commercial developers who were more apt to get jumpy when discussing free beer. However, now "free beer" may be the more appealing argument to make. The global economy is forcing cost considerations in everything people do.

With the vast inventory of open source software that now exists, there is a much better chance that the exact "itch" you have will already have been "scratched" by another very talented developer!

Thus, the freedoms of open source are still necessitated by the need to change/maintain underlying code, but there is heightened urgency to benefit from the free price as well.

Computing on the Edge

Today there are more network-connected points than ever before. These connections may be close by or distant. Linking remote corners of the earth creates tremendous opportunity for meaningful global conversations. Network growth could explode as netbooks, smartphones, and other highly mobile devices living "at the edge" are connected. At the same time, we'll see a vast increase in the number of

Chapter 12: Mobile Directions

connections any device can make on an ad hoc basis. Connections to local and distant computers will skyrocket as this kind of computing becomes cheaper and more powerful to deploy.

Here is one example of an innovative new product that could redefine where the edges are. In March 2009, Marvell introduced the SheevaPlug shown in Figure 12-4 — a $99 ARM-compliant little powerhouse with a developer toolkit. With this product, Marvell could capture the interest of hobbyists and developers, while encouraging innovative applications that might someday drive demand for Marvell's embedded processors and wireless chips. The SheevaPlug uses a standard wall socket to run at 1.2GHz, with 512MB Flash, 512MB DDR2, USB 2.0, and Gigabit Ethernet. It all fits into a package about the size of a wallet (though a bit thicker than mine!) The whole package operates on less than 5 watts, supports Ubuntu, and provides a developer toolkit. It's very easy to move around and repurpose.

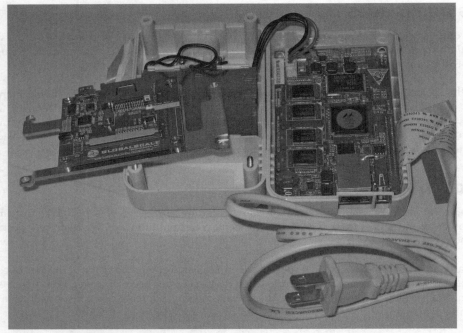

Figure 12-4

Initially, the SheevaPlug was designed for applications such as media servers and backup machines, both of which could tie into cloud computing.

With SheevaPlug, Marvell is taking an open source approach — making its designs available as open source and encouraging development activity at websites such as `http://plugcomputer.org`. Third-party developers have already incorporated these designs into their own products and systems, such as `http://pogoplug.com`.

Chapter 12: Mobile Directions

The Future

According to the December 2008 Pew Internet & American Life study, "More than three-quarters of the expert respondents (77%) agreed ... that the mobile computing device — with more significant computing power in 2020 — will be the primary Internet communications platform for a majority of people across the world." Does that projection ring true with you?

I conducted my own informal poll of friends and college students. In particular, I wanted to learn their thoughts on the future of Linux. While my survey isn't statistically significant, it's a start.

It quickly becomes clear that Linux faces hurdles to wider adoption. One friend put it this way:

> *There is a term used in economics called "focal points." It refers to the tendency for development in a given area to cluster around a few points (from among a multitude of equally likely points). It's why "things happen" on Mac OS and Windows. Things happen on the Mac because all the cool kids hang out there and things happen on Windows because everyone else is there. Actually, it's even simpler than that — "things happen there because they happen there." The main people on Linux are developers.*

Another hurdle is the preponderance of named distribution (choice is good but too much is confusing):

> *It's the opposite of the problem I confronted when coming to New England (where no intersections have street signs). With Linux, there are too many signs! "I thought you said this was Linux (or SUSE, or Ubuntu, or Red Hat, or Debian, or BSD, or Gentoo, or Knoppix, or whatever. Life's too short)."*

Another friend, who is a long-time Windows and Mac user, knows what goes into his computing gear:

> *Yes, of course I knew that OS X uses the BSD kernel. But it's not Linux. It's OS X.*
>
> *Can you run Mac software on Linux? No. Why? Because it's not Linux. Is Mac software part of the powerful Apple ecosystem? Yes. Can Linux take advantage of this powerful ecosystem? No.*
>
> *I am not using UNIX on my Touch. I am using OS X. A derivative of UNIX yes, but neither Linux or UNIX. A friend of mine in college had an Avanti. This was a Studerbaker fiberglass body with a Corvette engine. Was it a Corvette? No, it was an Avanti.*

Once upon a time, the great innovations might have come from corporate think tanks such as Xerox PARC (the Macintosh User Interface), but today these ideas come more often from a college campus. Consider these success stories, which grew out of college activities: Netscape, Google, Microsoft, and Linux. Slackware, the oldest Linux distro still under active development, was created by Patrick Volkerding as an outgrowth of a college project.

What happens on campuses will dictate what happens in future markets. I asked around MIT to get a sense of how students use mobile technology.

Even though WiFi is freely available for all MIT students, and virtually everyone has a notebook with WiFi, one student estimated that about 40 percent of students also have a phone with a dataplan:

> *There are a lot of iPhone owners around campus. Texting, e-mail, and calling, in probably about that order. Music and media if you have an iPhone or something that's specifically targeted at mobile media. I think that here people are tech-savvy enough that they probably use their mobile device for all of those things at least occasionally.*

253

Chapter 12: Mobile Directions

For students who do adopt Linux, varying combinations of these four reasons seem to explain why:

1. Free, as in Freedom (ability to modify/contribute)
2. Free, as in Free lunch (money saver)
3. Power/capabilities (including customization and development)
4. Reliability/security, and so on

A recent grad from Princeton shared his perspective: "All I know about Linux is that it was developed by a guy named Linus. I have no idea what kind of OS is running my iPhone, but I know I LOVE it!" His estimate was that 5 percent of students would know what a Linux distribution is.

At Olin school of Engineering, students are issued a notebook computer running Windows XP. As one engineering student observed, "A lot of non-techies are afraid of the (god-forsaken) terminal and the general lack of in-person support for Ubuntu/Fedora."

Taken to the extreme, the fear of a terminal program can be incapacitating! One Wisconsin student dropped out of college after her computer was configured with Ubuntu rather than Windows. She ordered a Dell laptop, expecting to receive a familiar Windows system. Instead, it arrived with Ubuntu installed. She called customer support to complain it wasn't what she ordered: "The person I was talking to said Ubuntu was great, college students loved it, it was compatible with everything I needed." She stuck with Ubuntu and floundered. Unable to connect to the Internet, or process required documents, she dropped out of the online program.

We can carry a number of lessons into the future:

- There's a good chance the next big breakthrough in mobile computing software will come from a college campus. Projects such as Google's Summer of Code encourage this. It will still take a special student (like Torvalds) to make it happen.
- If your grandma says she'd like a phone with bigger buttons, you'd better buy it for her! Appropriate technology is the best answer if the desire is to have people get things done.
- Linux, Ubuntu, and related distributions will only succeed in mobile devices on a large consumer scale if they are easy to use and well suited to the use-case scenarios.

Ubuntu, Linux, and Mobile Computing

How will Ubuntu Mobile evolve? One expert who would have a good idea is Mark Shuttleworth. He shared his thoughts on this topic at the Argentina Debconf8 in August 2008. A recorded interview, conducted by Barton George, then with Sun, offers quite a few insights.

With respect to netbooks, Shuttleworth sees part of their success in how they have been used:

> Built into the use case (of netbooks) is an implicit assumption that you're not going to try to use it in all the ways that people have traditionally tried to use their PCs. That makes (people) more willing to consider alternatives ... So it's a very exciting development for the Linux community ... the real willingness of people to buy a machine that doesn't have Windows on it but that delivers the things that Linux delivers really well ... great web experience, good security, efficient performance, lightweight hardware, and great economics.

Chapter 12: Mobile Directions

Shuttleworth also shared his thoughts on where the mobile industry is headed:

I think the mobile space is enormously productive for Linux. We're already seeing companies like Motorola quoting figures like 60% of their device market share going to Linux. . . . At the distro level and at the app-stack-framework level, the picture is much less clear.

We're quite strongly aligned with Intel around their Moblin initiative for mobile because they have proven open to participation. I think the key is figuring out how to bring together operators, bring together handset manufacturers, bring together chip manufacturers. I'm very hopeful Moblin will achieve that.

At this stage, I don't think anybody has definitively gained a level of critical mass. There's still a certain amount of alignment of the platforms to vendors. In the case of Moblin, it's strongly aligned with Intel. I think it would be very healthy to see broader adoption of that. You have Android from Google. For them to be successful they'll need to demonstrate broad participation. LIMO is very strongly aligned with Motorola, but increasingly an open environment as well.

The things I'd be looking for are an emerging consensus across chip manufacturers, handset manufacturers, and operators on how they want this thing (mobile) to work.

Open source or free software really is the right way to bring together an industry that is particularly fractious. I'm very hopeful that Linux will play out well but exactly how that will play I'm not sure.

Linus Torvalds made this observation:

It's a huge job to do a distribution. The reason there are hundreds is it is easy to start your own, but if you want to be a leader and introduce new code, the testing and Q&A involved is enormous. It depends on having enough users that you get coverage and it is unreasonable to expect too many large distributions. Ubuntu grew surprisingly quickly and maybe that can happen again. Your working knowledge of Ubuntu, Linux, and mobile technologies should be durable experience you can carry with you no matter how this change takes place. As the PC shrinks in size, it is on a collision course with the multifunction cellphone. Many expect the resulting impact to transform both devices and all the companies that make them. The new smartphones, always-on portable Internet devices that are part cellphone, part computer, change the rules of the game in computing because computing speed — at which Intel excelled — is no longer the most important factor. For a cellphone relying on a small battery, how efficiently a chip uses power becomes more important. . . . Cellphones outsell PCs by about five to one.

Summary

A good piece of advice is to "Expect the unexpected." An avalanche of exciting (and mostly unexpected) developments in hardware and software hit the market in 2007 and 2008. This was welcomed by a nearly overwhelming demand for small, powerful, mobile computing devices. The combination of avalanche and demand resulted in record revenues despite tough economic times. Demand for "Professional Ubuntu Mobile Development" should keep growing for many years into the future

13

Common Problems and Possible Solutions

This chapter attempts to gather several of the most common problems that have been posted to the ubuntu-mobile mailing lists as well as solutions that have been offered around the Internet.

The Boot Process Stops

Q: I downloaded Ubuntu MID, I extracted the files in the image, and I prepared a USB stick for booting with this image.

After the USB boots, I press Enter to start the installation and, after several OS messages, the boot process stops. The last message I receive is

```
will mount root from /dev/sdb
```

That's all.

A: This can be caused by improperly writing to the USB stick.

To copy the image to the USB stick (assuming /dev/sdb for the stick), use the following

```
$ sudo bs=1024 if=<image file> of=/dev/sdb
```

dd is a program used for low-level copying and conversion of raw data.

Chapter 13: Common Problems and Possible Solutions

Writing an image to a USB stick is like filling out a grid on a checkerboard. If you don't give it the 1,024 block size parameter, it writes one way (row-by-row), but with bs=1024, it writes column-by-column. When testing this, there was an approximate 40 percent success rate with just dd if=<image> of=<drive>, but with bs=1024, this was boosted to nearly 99 percent.

Application Icon Does Not Appear

Q: My application icon does not show up on the desktop.

A: To have an icon appear on the desktop, follow these steps:

1. Install a .desktop file into /usr/share/applications.
2. Install an icon into /usr/share/icons/hicolor/<size>/<type>

(for example, /usr/share/icons/hicolor/64x64/apps/myapp).

If Step 1 is done correctly, the application should show up in the UI.

If Steps 1 and 2 are done, it will also have the right icon.

If the icon is still not appearing, try adding the following:

    OnlyShowIn=GNOME;Mobile;

to the desktop file and the icon will appear on the desktop.

For example, cheese does not have OnlyShowIn, but it appears in Hildon desktop.

Hhildon desktop is the primary UI component of of the Hildon Application Framework for mobile devices.

Others also do not have OnlyShowIn but do not appear. There is a special implementation for Ubuntu Mobile, which is seen in the GConf keys:

    /desktop/hildon/htmlhomeplugin/onlyshowin_filter

Check the OnlyShowIn value in .desktop. If it is _False_, every desktop is shown. This command:

    /onlyshowin_ignore

always shows these apps, even if they don't have OnlyShowIn.

For example, cheese is in this list and is always shown.

Apparently this key should go away when every application is a good citizen and complies with the freedesktop.org standards.

Chapter 13: Common Problems and Possible Solutions

In addition, some utilities are available when dealing with .desktop files in the package desktop-file-utils. They are:

- `update-desktop-database` — Update the desktop-MIME mapping.
- `desktop-file-validate` — Validate a desktop file.
- `desktop-file-install` — Install a desktop file

Use them like this:

```
$ desktop-file-validate hello-world.desktop
```

Performing a Dual Boot

Q: Would it be possible to dual boot Ubuntu Netbook Remix on a Windows Mobile device?

A: This very much depends on the device hardware. A fair number of existing models support some form of alternate booting, either directly from some physical device, or through an application that replaces the currently running operating system with that from a physical device or selected file.

The majority of current devices supported by Ubuntu Mobile fall into the first category, and allow booting from alternate physical devices (most commonly USB keys, as used when testing).

It is entirely possible, if the hardware is sufficient, to support dual booting. Many devices are targeted for the low cost market, and to meet that, they incorporate a small SSD for data storage (4–16GB). Systems such as the Eee PC from Asus are capable of running both Windows and Linux, but not from the same drive because of space. It is possible to install Linux on a USB drive and run it "Live" from USB, and use the SSD for XP.

A useful application for this is GNU Haret, which you can find at http://handhelds.org/~koconnor/haret/ or http://handhelds.org/moin/moin.cgi/HaRET.

Setting a Flag Automatically

Q: Is there a way to set up configure.ac so that it automatically sets the `USE_HILDON` flag if building for the LPIA architecture?

A: It is possible to add the following rule to debian/rules:

```
ifeq ($(DEB_BUILD_ARCH), lpia):

DEB_CONFIGURE_EXTRA_FLAGS = --enable-hildon

endif
```

This causes `configure` to be called with an extra `--enable-hildon` argument when building on the LPIA architecture.

Chapter 13: Common Problems and Possible Solutions

It is also possible to add into `configure.ac` or `configure.in`, as follows:

```
hildon=false

HILDON_CFLAGS=""

HILDON_LIBS=""

AC_ARG_ENABLE(hildon,AS_HELP_STRING([--enable-hildon],[Turn on hildon

support]),[

    if test "x$enableval" = "xyes"; then

        hildon=true

        PKG_CHECK_MODULES(HILDON,[hildon-1],

                        HAVE_HILDON=yes,HAVE_HILDON=no)

        HILDON_CFLAGS="$HILDON_CFLAGS -DWITH_HILDON=1"

        PACKAGE_CFLAGS="$PACKAGE_CFLAGS $HILDON_CFLAGS"

        PACKAGE_LIBS="$PACKAGE_LIBS $HILDON_LIBS"

    fi

])
```

This will enable a `HAVE_HILDON` flag you can test at build time to enable or disable changes for the LPIA architecture.

Using USB

Q: I use Windows Vista and I want to prepare a USB stick to install ubuntu-mobile.img on my Samsung Q1 Ultra. What is the best way to do this?

A: The best way to do this is to use dd for Windows, which is available at http://uranus.it.swin.edu.au/~jn/linux/rawwrite/dd-old.htm.

Running Ubuntu on Freerunner

Q: Any chance that Ubuntu MID will ever be running on something like the Neo Freerunner? The Freerunner already has a Debian port so it might be possible?

A: There is nothing to prevent this, but it's not currently a target of the Ubuntu Mobile team to support phones: Ubuntu MID is about "MIDs," and mostly about an Internet experience. However, it's true that some recent MIDS are getting SIM cards and phone capabilities.

Chapter 13: Common Problems and Possible Solutions

Running Ubuntu on Arima

Q: As a related question, what would be the difficulties of getting Ubuntu MID running on, for example, the Arima UM650? Is Ubuntu Mobile hardware-specific?

A: Ubuntu Mobile with the Jaunty release has Arima support. Getting Ubuntu Mobile to work on other hardware is just a matter of getting the right kernel working. The linux-lpiacompat kernel ought to work on i586+ machines. The C7-M, for example, is one of the processors for which testing has been limited, but is of some interest as it has been deployed in a wide number of devices.

Ubuntu Intrepid UMPC Project

Q: I tried the Ubuntu Intrepid UMPC image on a 10-inch tablet PC and I really liked it. However, I can't seem to find enough official documentation about this project, so I was wondering if the project is still ongoing?

A: After the initial excitement about the release, few people have actually been committing their changes to the repositories; at least, the UMPC image is not listed as a target release with 9.04. That being said, there's no good reason why someone couldn't adjust the Intrepid ubuntu-mobile-default-settings package to work with 9.04, or continue development. The changes are mostly adjusting settings, such as GConf or something theme-related.

Installing Ubuntu Netbook Remix on a UMPC

Q: I just bought a device called a UMPC, can Ubuntu Netbook Remix be installed on UMPCs?

A: Yes, it can be installed and works well. It helps if you increase the panel size for touch and there's also a GConf key that increases the button/widget sizing in the launcher. It's easier to use with touchscreens (needs a restart of the launcher):

```
$ gconftool-2 --set /apps/netbook-launcher/tablet_mode --type BOOL true
```

Another useful tip is to change `$your_fav_gtk_theme`'s `gtkrc` to increase the `xpadding` and `ypadding` attributes for scrollbars, checkboxes, and radio buttons to make GNOME a bit more touch-friendly.

Using apt

Q: I tried to install a package using apt but I got a message that it was not possible to install as a result of broken dependencies.

Chapter 13: Common Problems and Possible Solutions

A: Try changing the source of your apt repositories. It can happen sometimes that they become out of sync and changing the location of the apt repositories and running an apt-get update can solve this problem.

Joining the Ubuntu Mobile Developers Team

Q: How can I join the Ubuntu Mobile Developers team?

A: If you need write access to the ubuntu-mobile repositories, join the Ubuntu Mobile and Embedded developers team on launchpad.net. This is a moderated group. However, a "sustained" contribution to the team should be enough to get you accepted. This can be anything from code, documentation, and artwork to advocacy work.

Using KVM or QEMU

Q: I cannot get either KVM or QEMU to work. What can I do?

A: If this happens to you, make sure that you have the KVM and QEMU packages installed. Next, download an older hardy release of MID from http://cdimage.ubuntu.com/mobile/releases/hardy/mid-8.04.1-kvm.tar.gz.

This package includes both a virtual image called root.qcow2 (qcow2 is a QEMU disk image format) and a launching script called ubuntu.kvm:

```
#!/bin/sh

kvm -soundhw all -m 128  -hda root.qcow2 "$@"
```

Extract the downloaded package

```
$ tar -xzvf mid-8.04.1-kvm.tar.gz
```

and then move into the new directory and execute the script

```
$ cd ume-8.04-kvm

$ sudo ./ubuntu.kvm
```

Graphical Corruption

Q: When I install on a device I see the status bar and the installation seems fine, but when the desktop should appear, it looks like the device has crashed — all I see is graphical corruption.

A: Start Ubuntu in Safe Graphics mode. This will start Ubuntu using Vesa graphics drivers, which are compatible with practically every graphics card made within the last ten years.

Poor Performance

Q: I installed Ubuntu Netbook Remix on an Eee PC and the performance is very poor. I installed the default Ubuntu Jaunty to test and performance was much improved.

A: This is due to a video driver bug, which means that tiling is not enabled by default (this is essential for Ubuntu Netbook Remix as the launcher is based on GL). A fix has been made and can be found at `http://people.ubuntu.com/~apw/lp349314-jaunty/`.

The files to install are linux-headers-2.6.28-11-generic_2.6.28-11.43~lp349314apw5_i386.deb and linux-image-2.6.28-11-generic_2.6.28-11.43~lp349314apw5_i386.deb. Download both of them into a folder and install them by running the following:

```
$ sudo dpkg -i *.deb
```

This information may well be wrong by the time this book is published. Please check on the mailing list for the latest details.

Summary

These are common problems and questions relating to Ubuntu Mobile that pop up frequently on the IRC channels and the mailing list. If you have questions, you can always ask on irc.freenode.net in the channel #ubuntu-mobile or send an e-mail to the mailing list.

Ubuntu's Right ARM

Now that Ubuntu has been released for the ARM (Advanced RISC Machines) platform, there are many intriguing new application possibilities. An entire class of solutions is available in the realm of power-efficient computing.

One example is the SheevaPlug. In February 2009, Marvell brought to market a new device called the SheevaPlug, based on its ARM processor. Marvell provided an open architectural blueprint for "plugs" reminiscent of the early, standards-based component designs that laid the foundation of the PC industry. Marvell's combination of technology and product vision has popularized an entirely new market segment: plug computing. (Go to `http://plugcomputer.org/` for more details.)

The *Wall Street Journal* reported on the launch of this new product and segment with an article on February 23, 2009. The coverage hinted at a new mini-industry:

> "There are multiple pain points we think we can solve here," said Simon Milner, a Marvell vice president and general manager of its enterprise-business unit.
>
> One big problem is consumers' growing hoards of digital photographs, music and videos — increasingly stored on laptop computers that could be lost or stolen. Companies already sell devices to safeguard such information, including devices that combine disk drives and networking features.
>
> Marvell says it believes plug computers can make those products more useful, a concept some companies first discussed at the Consumer Electronics Show in January without disclosing Marvell's involvement. The devices connect with an Ethernet cable to a home router and to storage devices using the widely employed technology called USB.

The Wall Street Journal picked up on the power savings potential — less than five watts for a SheevaPlug compared to the electrically thirsty alternatives. In addition, the Wall Street Journal mentioned $49 as an eventual price point for some models.

Appendix A: Ubuntu's Right ARM

Rob Enderle, a principal analyst and President for the Enderle Group, described the significance of this event: "Plug Computing is one of the most revolutionary technologies to come to market this decade. It represents a future where processing power is inexpensive and energy efficient founding an age where computing is something you don't worry about because it just works."

I'd Give My Right ARM for a SheevaPlug

I was one of the lucky ones to get my hands on a SheevaPlug soon after the initial announcement. Within the first week, they were sold out and units were on backorder. The initial demand far exceeded what Marvell had anticipated. It was one of the clear signs that this segment has "legs to run with." This is what you received (and didn't receive) with the initial $99 Developer Kit:

- Marvell Sheeva ARM CPU at 1.2 GHz
- 512MB DDR2 @ 400 MHz
- 512MB NAND Flash (direct boot from Flash)
- Gigabit Ethernet/RJ45
- USB 2.0 (direct boot from USB)
- SDIO
- JTAG/UART serial interface
- No fan. No drive. No noise.
- Elegant packaging/Built-in AC power supply
- Ubuntu pre-installed
- No license fees ("open" reference design)
- No graphics/VGA support
- No built-in WiFi

The power consumption varies depending on the tasks that are run, but it won't exceed 7 watts as you can see from the following power profiles:

- With cpuidle in C1, 2.5W
- With cpuidle in C0, 3.0W
- 3.3W at idle with gigabit ethernet
- 5.0W idle with gigabit+HDD USB+serial
- 7.0W + 100% cpu

Figure A-1 shows what a SheevaPlug looks like inside.

Appendix A: Ubuntu's Right ARM

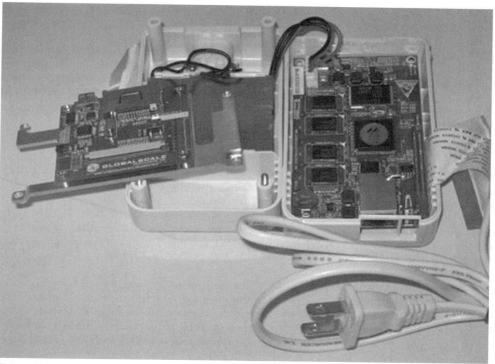

Figure A-1

Application Ideas

My initial goals for the SheevaPlug included several energy efficient, 24/7 applications: a backup server, a media server, as well as a home automation and information system. For software, I was looking forward to using the tools we are familiar with in Ubuntu: Python, Django, networking databases, a fully stocked application repository and more.

The SheevaPlug proved easy to work with for someone with basic Linux familiarity and prior experience working with Ubuntu. Although there is no graphics or VGA support, it was easy to SSH into the device or manage it via the JTAG serial connection. (MIMO Monitors is shipping 7-inch screens that work across a USB connection. Its USB to LCD controller makes it possible to display visuals without the need for VGA or video support. Key technology has recently been released as open source and Ubuntu support is possible.)

In the first days of experimenting with the SheevaPlug, I was able to successfully attach wireless USB 802.11g radios (from Edimax). I had the same success attaching a USB webcam to the SheevaPlug. Before I knew it, I had all the core ingredients of a home control/surveillance system powered by the SheevaPlug.

I was both thrilled, and a little startled, when I discovered I could easily run simple OpenCV applications on the SheevaPlug.

Appendix A: Ubuntu's Right ARM

OpenCV (Open Source Computer Vision) is a library of programming functions that are aimed mainly at real-time computer vision.

Although OpenCV has been in open source for a long time, its website doesn't claim support for the ARM platform. There are dozens of dependencies required to make OpenCV work with Python. Plus, there is typically heavy dependence on floating point hardware for graphics processing (this version of the SheevaPlug lacks floating point hardware). Even these early days of running Ubuntu on ARM are proving productive: OpenCV ran fine on the SheevaPlug.

Relocatables

Another application area of interest is in a market segment I call "Relocatables" — as a short form of "Relocatable Computing." The essence of Relocatables is that you can move your computer infrastructure easily from place to place with minimal setup.

Relocatables would be useful if your network moves, such as to cover outdoor sporting events or disaster recovery locations. Anywhere network infrastructure runs the risk of being damaged or destroyed, Relocatables could be a useful solution.

One simple example of this is municipal WiFi, where wireless devices might need to be moved in support of community events. It turns out that one of the toughest problems with "Municipal WiFi" is finding electricity to run even low-powered radios and computers. Often, building owners will provide access to a roof for placing antennas, but power is absent. Thus, we turned to solar as a workable solution for providing power. The SheevaPlug is low power, and now well within reach of being powered by solar.

As a result, the challenge was to step it up a notch. So, a challenge goal was put forward to build a cluster of SheevaPlugs into a solar-powered super computer. This is what came to be known as SWARM.

What's the Buzz on SWARM?

SWARM is an acronym that stands for:

S — Sheeva/Solar-powered

W — Wired/Wireless/Watt-Friendly/Web-Server

A — Advanced/Application

R — Running/Racing

M — Memcached/Machines

My goal was to gather the needed parts and assemble a SWARM on June 17, 2009, for the fifteenth anniversary of the Boston Linux User (BLU) Group. As you can see from the following photos, which recap the highlights in our construction of the SWARM, everything worked as planned.

Appendix A: Ubuntu's Right ARM

Marvell generously contributed ten SheevaPlugs in support of this early, open innovation project. Our original hardware design for SWARM was inspired by the battle cry, "Stack it, don't rack it!" (See Figure A-2.)

Figure A-2

On June 16, 2009, we spent a day configuring and testing a dozen SheevaPlugs. We installed the latest version of the Ubuntu operating system. Rabeeh Khoury, a Linux and SheevaPlug expert from Marvell, contributed his expert knowledge and assistance. (Rabeeh is the author of the SheevaPlug installer. He helped configure the SheevaPlugs so that the Linux kernel was stored on the NAND Flash and the rest of the Ubuntu system was installed on fast 8GB SDIO cards.) Our initial configuration was more horizontal than vertical, with a simple table layout of devices, as shown in Figure A-3.

Appendix A: Ubuntu's Right ARM

Figure A-3

For unit testing of each plug, we exercised basic Ubuntu software (e.g., Python). We then installed a variety of additional Ubuntu packages (such as OpenCV and MEMCACHED.) We planned to use MEMCACHED to build a super web cluster. There are vast quantities of web servers burning up enormous amounts of electricity on a 24/7 basis. A device similar to the SheevaPlug, or a cluster such as SWARM, could potentially save considerable energy while providing acceptable web server performance.

We initially connected our SheevaPlugs together using two GigaBit switches. (The D-Link DGS-2208 10/100/1000 Mbps 8-Port Desktop Green Ethernet Switch is smart about how it consumes electricity — it lowers electrical consumption when ports are inactive, thereby saving power and reducing heat.) We would later scale up to a bigger switch.

A Solar-Powered Barn Raising

Kurt, a grad student at MIT and a guru in solar, municipal wireless, and cluster computing provided inspiration for this project work. As he once observed, "More of cheap trumps fewer of better." With his help, we evaluated some of the ideas in FAWN ("Fast Array of Wimpy Nodes") and also considered a number of other energy efficient platforms (such as OpenWRT and the WGT-634U from Netgear).

Then we needed to figure out how to build this — after all, a cluster required a number of technical skills. We turned to history for an organizational model.

In North America during the eighteenth and nineteenth centuries, communities came together to help families build barns. Virtually every farming family needed a barn, yet it would take too long and was

Appendix A: Ubuntu's Right ARM

impossibly difficult for any one family to do this alone. Barn raising required planning and preparation and organization of various skills and materials — with participants volunteering their time and skills for free. The barn raising was usually carried out with a sense of urgency — families needed to use their barns! We took a barn raising approach in building our SWARM.

We gave advance notice at the May BLU meeting and set the date for June 17. I brought along a bin full of ready-made SheevaPlugs. So you might translate this into a "bin raising," as shown in Figure A-4.

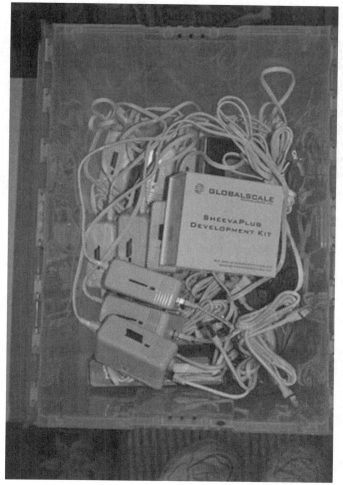

Figure A-4

About ten people met at the MIT building by 1:30 P.M. Jerry Feldman, a long time leader of the BLU group (along with John Abreau) marshaled the assembled resources into a team that is shown in Figure A-5.

Appendix A: Ubuntu's Right ARM

Figure A-5

Sage Radachowsky brought his solar panel, charge controllers, battery, and pliers. Within an hour, he had everything wired up, providing sufficient solar power to keep our ten SheevaPlugs and the power hungry switch running smoothly.

Bill Bogstad, a Boston Linux User Group attendee, took responsibility to network the SheevaPlugs with a gigantic gigabit switch. The team made sure the WAAV (packed inside a standard backpack, with the pole sticking out) was working. This wireless/cellular adaptor was quickly operational, plugged into the Gigabit switch, and acted as a DHCP server to provide network addresses for each of the SheevaPlugs (see Figure A-6).

Appendix A: Ubuntu's Right ARM

Figure A-6

Dave ducked inside MIT with an Acer Aspire netbook running Ubuntu to tune the SWARM (see Figure A-7).

Figure A-7

Appendix A: Ubuntu's Right ARM

Soon Dave had logged in and was administering the SWARM while running primitive performance tests. All of his system management was done wirelessly from an inexpensive Acer Aspire netbook. Remarkably, the SWARM was connected to the world without the aid of network or power wires.

The collective "bin raising" team brought the SWARM to life and connected it to the Internet and MIT's internal network in about three hours. Jerry and others watched in the sunshine. Figure A-8 shows how the SWARM electrically and logically came to life.

Figure A-8

The magic of the afternoon could rival the best illusions of David Copperfield in Las Vegas. But every bit of this was real.

What Did We Learn?

Later that evening, we "relocated" the system indoors. Lacking solar power, we just plugged it into MIT's wall plugs. The entire ten-node system, including switch, consumed a mere 70 watts. Yet it had enough power and network cables for us to make the obvious comparison to spaghetti code — this was a "spaghetti cluster!" (See Figure A-9.)

Appendix A: Ubuntu's Right ARM

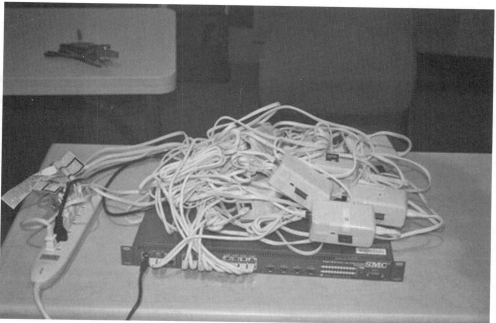

Figure A-9

The BLU meeting that evening was attended by some 60 people. As one prominent developer in the software industry entered the meeting room, he pulled out his pocket computer and commented to me "I'm already connected to the cluster. What's it do?"

Everyone saw the demonstration and asked plenty of questions. (Can you run it on USB power?) Bill Bogstad played a key role in assembling the SWARM that afternoon and he also sat through the presentations we delivered that night. He agreed to share what he took away as lessons learned:

- ❏ Low-power Linux-based computer devices can really push bits (and likely compute them as well)!
- ❏ Low power can *really* mean low power (<100 watts for a *cluster*)!
- ❏ By implication, low-end cluster computing can do real work.

With this meeting, we demonstrated the potential for low-power ARM computing in "Relocatable" scenarios.

The versatility of Ubuntu — both the vast archive of software and the comprehensive coverage of platforms — underscores how truly valuable this platform is for developers and end users. Ubuntu is a superb platform for experimenting, exploring, and doing real work at the cutting edge of technology. With the ARM release, energy efficient computing has arrived. One new solution segment is plug computing and another segment worth considering is "Relocatables."

Appendix A: Ubuntu's Right ARM

A handful of meeting attendees, some of whom helped build the SWARM and others who have participated in the discussion — the *barn raisers* and *barn stormers* — gathered for a photo that's shown in Figure A-10.

Figure A-10

I've now worked with Ubuntu and ARM for about four months. It seems clear to me that the combination of power, ease of use, and compatibility makes Ubuntu right for ARM.

Git Usage

Git is a distributed SCM (Source Code Management) System that was written by Linus Torvalds in 2005 to follow the development of the Linux kernel. Since then, it has been developed by a rather large group of hackers around the world.

Intended to be used in a distributed environment, `git` is really good for sharing work in a group of developers no matter what the group's size or geographical location. It's optimized for merging code from other developers into your repository.

So that you can understand `git` as quickly as possible, we have organized this appendix according to the kind of project for which it might be used. In each section, basic `git` commands are used, which are necessary for a developer to play a role in the project.

First, the basic repository commands from `git` — `clone`, `init`, `fsck`, and `gc` — are used, and then commands that are necessary to work with `git` as an individual developer in a standalone project are shown. Finally, commands that are useful for a multi-person project are demonstrated with different roles: contributor, maintainer, and integrator.

How Git Works

Git has three different objects that it uses to describe the whole project: `blob`, `tree`, and `changeset`. Each one of these objects is used for a different purpose, but all are involved in tracking the content modifications in your project.

Blob Object

A `blob` object is just a binary `blob` of data and doesn't refer to anything. The only data verification done in the `blob` object is that it is indexed by its SHA1 hash and, besides that, has no other attributes — no name associations, no permissions. It's a pure `blob` of data that can be translated into *file contents*.

Appendix B: Git Usage

Tree Object

The `tree` object is defined by a list of permission, name, and `blob` data sorted by name. This means that two identical but separate `tree` objects will share (or point) to the same `blobs`. This gives you a really interesting feature in git for tracking file renames. We don't really mess with the data anyhow, just point to it via a different path — or a different `tree` object.

Changeset Object

The `changeset` object introduces the notion of history into a git repository. It doesn't merely give you the physical state of the `tree`, but also describes why and how you got to that particular state. A `changeset` is defined by the `tree` that it results in, its parent `changesets` (zero or more) that took the source code up to that point, and a human readable comment on what happened.

How Git Objects Relate to Each Other

It's simple to understand git if you understand how these three types of objects relate to one another. Figure B-1 shows how git works and how it tracks the contents of your source code.

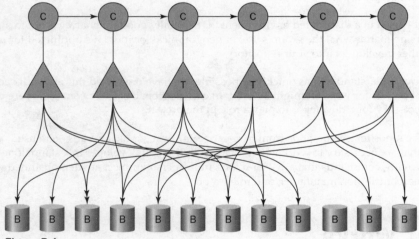

Figure B-1

Basic Repository Commands

Before you can actually start using git, you need to have a repository to work with. If you're starting a standalone project, you will initialize an empty repository and add source code to it as you develop; but on the other hand, if you're working on a pre-existing repository, you need to clone that in order to have a local copy to work with.

Appendix B: Git Usage

Also, it's important to keep your repository clean and free of errors so you don't lose data while working on it. The last two commands in this section are useful for housekeeping your git repository.

Creating a New Repository

If you want to start a new project from scratch, you use the git command `init`. This command creates a `.git` directory in the directory you are in. That will be the start of a project.

```
$ mkdir -p projects/my_first_project
$ cd projects/my_first_project
$ git init
Initialized empty Git repository in /home/user/projects/my_first_project/.git/
```

If you take a look at the .git directory, you will see a bunch of directories and files that git uses to actually track the content of your project. Yes, that's right, git tracks content only, not files. This means that when you move or rename a file, for git it doesn't really matter — it points to the same content via a different path and that's it, or, in other words, the `tree` object has a different way to reach the `blob` object.

Cloning an Existing Repository

You might be trying to collaborate with an existing project. In that case, you would have to clone the main `tree` of that project and start working on top of that. In that situation you could use the following:

```
$ mkdir -p projects/
$ cd projects
$ git clone git://<url>/path/to/project.git
```

For a real-life example, you could clone the Linux kernel `tree` from Linus Torvalds with the following command:

```
$ git clone git://git.kernel.org/pub/scm/linux/kernel/git/torvalds/linux-2.6.git
```

Cloning will literally make a local copy of the remote repository, meaning you will have the complete history of that project available for you. You can work completely offline, commit to your local copy, generate *diffs*, check out older code, create branches, anything.

Because git's goal is to be a distributed SCM, it's just natural that you can do all that work on a local copy of the mainline remote repository. Without this *freedom*, it would be impossible to have a real distributed SCM.

Checking Repository for Errors

It seems almost inevitable that filesystems get corrupted. You can try to avoid such corruption by checking your repository's health with the following:

```
$ git fsck
```

Appendix B: Git Usage

This is a rather slow operation as it has to check whether you have unreachable objects, dangling objects amongst other common errors. Upon running this command, you might see something like the following:

```
dangling tree b47648f75aa0ae2a0c9508bdd6efe00bb2da17c6
dangling tree 2eedf3560f4d0d403b0047f5c8ebc5f163841e53
```

This means that if you have two `tree` objects dangling (or existing objects that aren't used) they could be removed from your repository. Remember that dangling objects are not a problem; they often appear in a repository, especially if you *rebase* your commits frequently.

Housekeeping a Repository

It's important to keep your repository clean; this will help git do its job faster. Here you look at how to remove loose objects, re-create git's packs, and remove the unnecessary ones. Start by running the following:

```
$ git gc
```

`gc` does its best to repack your repository, and remove loose objects and unused old packs. However, it's for sanity housekeeping only. `gc` doesn't try to be too aggressive by itself. That's when you might want to use the following three commands:

```
$ git repack
```

With proper options, you can tell `git repack` to get all your objects and, instead of creating an incremental pack with the delta, create only one pack with all our objects. This will save a considerable amount of disk space and help git work a bit faster. To achieve this, you can use `git repack -a -f -d`.

The command `git-prune` will prune (remove) all unused loose objects from your repository. In other words, all loose objects already contained in a pack will be purged.

```
$ git prune
```

Similar to `git prune`, the command `git prune-packed` will remove unused packs from your git repository.

```
$ git prune-packed
```

There's not much to say about this command. You can use `-n` to just check the packs that will be removed and `-q` to suppress the progress indicator.

Individual Developer

Now that you have your set of basic commands, you can start doing some real work with git. At first, let's see what a developer working alone in a standalone project would have to know in order to track his project's history with git. To achieve that, first check which branch you are working on with the following command:

Appendix B: Git Usage

```
$ git show-branch
```

This command will basically show you where you are. In other words, it will print the branch and commit names where you are currently located. In a multi-branch repository, it will show a `tree` of how the branches are related. For example, it could show something like the following:

```
* [master] commit name 1
 ! [topic] commit name 2
  ! [topic2] commit name 3
---
*++ [master] commit name 1
```

```
$ git-log
```

The preceding code will show the complete log of the repository, generally using a pager such as *less*, so that you can take a look at the project's history `changeset` by `changeset`.

```
commit 172ee4898de698d164e902c96efce3e0981707c0
Author: Felipe Balbi <my_email@mydomain.com>
Date:   Fri Feb 27 22:02:33 2009 +0200

    commit 1

    Long commit description

    Signed-off-by: Felipe Balbi <my_email@mydomain.com>

commit 09e9f2e9b5335489f1a1c46a846ad3f649bd7c3c
Author: Felipe Balbi <my_email@mydomain.com>
Date:   Fri Feb 27 01:22:45 2009 +0200

    commit 2

     Long commit description

    Signed-off-by: Felipe Balbi <my_email@mydomain.com>


commit 4aed4f88d233df05adf6ed3135a0bdbbccf5f21b
Author: Felipe Balbi <my_email@mydomain.com>
Date:   Wed Feb 25 21:50:39 2009 +0200
```

Appendix B: Git Usage

```
    commit 3

     Long commit description

    Signed-off-by: Felipe Balbi <my_email@mydomain.com>
$ git branch
```

This command can be used to see the current branch you're in, to create a new branch, to delete a branch, and to rename a branch. Let's see how.

$ git branch without any arguments will show the branch listing. The branch marked with * is the one you're working in right now:

```
$ git branch

* master

  topic1
  topic2
```

$ git branch *branch_name* will create a new branch called *branch_name*:

```
$ git branch topic3
```

The results are as follows:

```
$ git branch

* master

  topic1

  topic2
  topic3
```

$ git branch *-m new_name* will rename your current branch to *new_name*:

```
$ git branch -m master_old

$ git branch

* master_old

  topic1

  topic2

  topic3
```

Appendix B: Git Usage

`$ git branch -D` *branch_name* will delete *branch_name*:

```
$ git branch -D topic3

Deleted branch topic3 (172ee48).

$ git branch

* master

  topic1

  topic2
```

The main purpose for `$ git checkout` is to switch between branches. For example, `git checkout topic1` will move your current working `tree` to `topic1`:

```
$ git checkout topic1

Switched to branch "topic1"

$ git branch

  master

* topic1

  topic2
```

Using these commands gives results listed:

- ❏ `$ git add` allows you to add new files or modified files to the index and, thus, to the next commit.
- ❏ `$ git diff` generates a *diff* between any two valid git objects, generally two commit IDs. Also used to see the *diff* between the current working tree and the index.
- ❏ `$ git status` shows you what's different in your repository — new files, deleted files, modified files — but only before you commit those changes to the index.
- ❏ `$ git commit` advances the current branch with the changes added to the index.
- ❏ `$ git reset` is used to undo changes. In a modified repository, you can use `git reset -hard` to lose all uncommitted changes.

  `git reset` *will literally forget any uncommitted changes and you will be unable to get those back. Be careful when using it.*

- ❏ `$ git merge` merges two or more local branches based on a desired strategy, or the default one.
- ❏ `$ git rebase` is used to maintain `topic` branches, meaning that when you move `master` forward, you can keep changes to `topic current`.

Appendix B: Git Usage

- ❏ `$ git rebase master topic` is used to let Git re-apply `topic` on top of the current `master`, updating all of the `topic` history.

- ❏ `$ git tag` marks a known point in history. Generally used to make a release out of the current state of the `tree`.

Contributor

A contributor — a developer who contributes to somebody else's project — will need a different set of commands in order to get her work integrated in the upstream project. It's important here to know how to work with a shared repository, when to rebase, when to merge, and so on:

- ❏ `$ git pull` pulls new code from upstream.
- ❏ `$ git push` publishes code to the central repository if you have permission.
- ❏ `$ git format-patch` generates an mbox patch file that can be mailed using `git send-email`.
- ❏ `$ git send-email` sends the patches generated with `git format-patch`.

Integrator

The main developer of a shared project will generally play the role of integrator. When playing that role, you might have to integrate (from now on called *push*) another developer's work. It's important to remember that on a shared repository you should not rewrite the history of the repository. Otherwise, everybody working downstream will not be able to pull again from your shared `tree`, forcing them to clone the `tree` again.

- ❏ `$ git am` is used to apply patches in *mbox* format mailed to you by your contributors.
- ❏ `$ git pull` will fetch and merge code from your trusted contributors.
- ❏ `$ git revert` is used to undo broken or unneeded `commits.u`.
- ❏ `$ git push` is used to publish code to a central and shared repository.

Maintainer

Being a maintainer is generally the same as being an integrator. On a few projects, however, that might become a completely different role. Imagine the maintainer being the main developer on a project and the integrator being the person doing some packaging work, for instance. In those cases, you have to play a different role with a different set of commands:

- $ `git pull-request` generates a *pull request* for the main developer of your project if you are maintaining only part of a big repository.
- $ `git merge` merges upstream code into your branch so you can keep your development going.
- $ `git mergetool` calls your preferred merge tool to help you solve the issues in case of merge conflicts.

Repository Configuration

In order to make the repository work more comfortable, it is possible to add some configuration variables, which can help you avoid some repetitive work. The following are the most common and necessary ones:

```
$ git config -global user.name "My Name"
$ git config -global user.email "my_mail@mydomain.com"
$ git config -global color.ui auto
$ git config -global merge.tool kdiff3
$ git config -global sendemail.smtpserver smtp.domain.com
```

Repository Administration

If you are the repository administrator, you will need to be sure `git daemon` is running and your users have access only to the necessary resources on your server:

- $ `git daemon` is the daemon that is used to allow anonymous download from your repository.
- $ `git shell` is used as a restricted login shell for shared repository users. It will not allow access to the console, but it will allow developers to push and pull from the central repository.

Hosting Your Project on Launchpad

Launchpad is a hosting and collaboration platform for software development projects. It offers bug tracking, code hosting, translation services, and a Q&A service. It will host your project's source code using the Bazaar version control system.

To do this, register the branch at `https://code.launchpad.net/people/+me/+addbranch` (you need to be registered with Launchpad for this). Fill out the web form which registers the new branch to Launchpad.

It can also import CVS and SVN so existing projects have this resource available to them. Launchpad provides this free service and then keeps that trunk branch up-to-date. This allows you to make your own Bazaar branches from the project trunk, and keep them up-to-date by merging from trunk over time as you develop your features.

To request an import, make sure the project is registered on Launchpad, or register it yourself. Then visit the request page at `https://code.edge.launchpad.net/+code-imports/+new` and fill out the details.

This will do the following:

- Create an empty branch to contain the imported code.
- Subscribe you to it so that you will be notified both when the initial import completes and subsequent updates import new revisions.
- Notify the import operators who will check the import location and approve the import.

The initial import can take a long time — up to several days, depending on the number of revisions that need to be converted. Once the import is established it will be updated from the CVS or Subversion branch every 6–12 hours, although an import can be requested at any time by clicking the Import Now button on the import page.

Appendix C: Hosting Your Project on Launchpad

At the moment, the service is restricted to tracking only the main branch of each project.

If you are having trouble using the service, ask on IRC in the channel #launchpad on the network Freenode.

> *Internet Relay Chat (IRC) is a form of real-time messaging or synchronous conferencing. It is provided by clients such as Xchat (sudo apt-get install xchat).*

Using Bazaar and Launchpad

This requires that Launchpad know about your SSH key. A quick way to find out your ssh key is to run the following command:

```
$ cat ~/.ssh/id_rsa.pub
```

It outputs something like this:

```
ssh-rsa
AAAAB3NzaC1yc2EAAAABIwAAAQEA8gqBwjpIQ/FI2sxh7J4VOSuSKoIvdaIWsLA9o4YOBA/
8UFN4FOM4cSrUiOq0zT71hpGPo9980B5zXbgGg6b4H5nXx0MIctfGt0yQeTY4aGYhj97/W34r/ExculubXL
ACEZzZv1NZJfG3SnSEwFF1E90tAu1waltK+paiIli/ONWS2VVLHUBhEXDTpZk8RdVNqixaj08NTjJ
AoqNSG99FJxx0AGQQp7ZDih/9Y+ip+nhHVYyGHf3Z7JPccEuS/m15wg1OOHHv6nzgbxtQlVtFj
ZHPObRKbSm3Nk3cnrD/urfFYKYkCoLDdlzYnOA7NyHcUb9TkPVHS+SdElXZ5B2Wiw== ian@lawrence
```

This can be pasted into the Update SSH keys section of Launchpad.

If you do not have an SSH key, run the following command to create one:

```
$ ssh-keygen -t dsa
```

Bazaar itself is installed with the following command to create one:

```
$ sudo apt-get install bzr
```

The code that will be hosted in the following example will be the first version of the Ubuntu Mobile Guide — Ubuntu 7.10 (Gutsy Gibbon) Developer Alpha — which is available on http://umeguide.net/.

Start off by telling Bazaar about yourself:

```
$ bzr whoami 'Ian Lawrence root@ianlawrence.info'
```

Next, move into the folder with the code and initialize the local repository:

```
$ cd mobileguide
$ bzr init
```

Look at the "status" of Bazaar:

```
$ bzr status
```

Appendix C: Hosting Your Project on Launchpad

It shows something like this:

```
unknown:
  C/
  README
  images/
  mobileguide.pot
  validate.sh
```

Next, add these to the directory by running the following:

```
$ bzr add
```

The command recursively adds everything to the directory. Things can also be added individually by specifying the file after the `add` command. They can also be removed by specifying the file after the `remove` command:

```
$ bzr add README
$ bzr remove README
```

When the code is in a reasonable state for upload, the changes first need to be committed to the local repository before they can be uploaded to a remote server (in this case Launchpad). This can be confusing to new users.

```
$ bzr commit -m "Example repository for the Ubuntu Mobile Book"
```

Now the code can be uploaded. It is possible to tell Launchpad about yourself by doing something like this:

```
$ bzr launchpad-login ianlawrence
```

Run the following:

```
$ bzr push --use-existing-dir sftp://ianlawrence@bazaar.launchpad.net/~ianlawrence/+junk/mobileguide
```

In the URL, `ianlawrence` is my Launchpad user name. After the ~ you can then include either your Launchpad user name or a team name; `project` is the name of the project in the Launchpad URL (in this case `+junk` as the project does not exist on Launchpad) and `branch` is what you would like to call the branch — in this case `mobileguide`.

The code is available at `http://bazaar.launchpad.net/~ianlawrence/+junk/mobileguide/changes`.

Checking Out the Branch and Working on It

If you want to contribute to the developer alpha version of the guide, you can check out the code at
`$ bzr branch http://bazaar.launchpad.net/~ianlawrence/+junk/mobileguide/`.

Desktop Power Applet Code

The following desktop power applet is written in Python. It is dependent on the python-gnome2-extras package. Therefore, it needs to be installed using apt.

First, import the necessary libraries:

```
import os
import sys
import gtk
from gettext import gettext as _
from egg.trayicon import TrayIcon
```

Note that the native way to display an icon is to use `gtk.StatusIcon` (new in PyGTK 2.10) rather then `TrayIcon`. Next, create a class called `PowerTray` with a call to `super` so that methods in the parent class (`TrayIcon`) are available:

```
class PowerTray(TrayIcon):
    def __init__(self):
        super(PowerTray, self).__init__('power')
        self.icon_theme = gtk.IconTheme()
```

The main power management code for a logout on a MID device is as follows:

```
if result == gtk.RESPONSE_YES:
        os.system('pkill hildon-desktop')
    dialog.destroy()
```

The main power management code for a shutdown on a MID device is as follows:

```
if result == gtk.RESPONSE_YES:
        os.system('gdm-signal -h')
        os.system('pkill hildon-desktop')
    dialog.destroy()
```

Appendix D: Desktop Power Applet Code

The main power management code for a reboot on a MID device is as follows:

```python
if result == gtk.RESPONSE_YES:
        os.system('gdm-signal -r')
        os.system('pkill hildon-desktop')
    dialog.destroy()
```

Here is the full source code:

```python
#!/usr/bin/env python
# -*- coding: utf-8 -*-
# Copyright (C) 2008 Rodrigo Cesar Lopes Belem
# Author: Rodrigo Cesar Lopes Belem <rodrigo.belem@gmail.com>
#
# This program is free software; you can redistribute it and/or modify
# it under the terms of the GNU General Public License as published by
# the Free Software Foundation; either version 2 of the License, or
# (at your option) any later version.
#
# This program is distributed in the hope that it will be useful,
# but WITHOUT ANY WARRANTY; without even the implied warranty of
# MERCHANTABILITY or FITNESS FOR A PARTICULAR PURPOSE.  See the
# GNU General Public License for more details.
#
# You should have received a copy of the GNU General Public License
# along with this program; if not, write to the Free Software
# Foundation, Inc., 59 Temple Place, Suite 330, Boston, MA 02111-1307 USA
# depends python-gnome2-extras,
import os
import sys
import gtk
from gettext import gettext as _
from egg.trayicon import TrayIcon

class PowerTray(TrayIcon):
    def __init__(self):
        super(PowerTray, self).__init__('power')
        self.icon_theme = gtk.IconTheme()
        self.icon_theme.set_custom_theme('Human')

        self.eventbox = gtk.EventBox()
        self.eventbox.connect('button_press_event', self.on_button_press_event)
        pixbuf = self.icon_theme.load_icon('gnome-session-halt', 32, 0)
        self.tray_image = gtk.image_new_from_pixbuf(pixbuf)
        self.eventbox.add(self.tray_image)


        menu_ui = \
"""
<ui>
    <popup name="TrayIconMenu">
      <menuitem name="lock" action="lock"/>
      <separator/>
      <menuitem name="reboot" action="reboot"/>
      <menuitem name="shutdown" action="shutdown"/>
```

Appendix D: Desktop Power Applet Code

```
            <separator/>
            <menuitem name="logout" action="logout"/>
        </popup>
    </ui>
    """
            self.uimanager = gtk.UIManager()
            self.uimanager.add_ui_from_string(menu_ui)
            accelgroup = self.uimanager.get_accel_group()
            actiongroup = gtk.ActionGroup('Power')
            actiongroup.add_actions(
                            [
                    ('lock', gtk.STOCK_DIALOG_AUTHENTICATION , _('_Travar'), None,
                        _('Travar'), self.on_lock),
                    ('reboot', gtk.STOCK_MISSING_IMAGE , _('_Reiniciar'), None,
                        _('Reiniciar'), self.on_reboot),
                    ('shutdown', gtk.STOCK_MISSING_IMAGE , _('_Desligar'), None,
                        _('Desligar'), self.on_shutdown),
                    ('logout', gtk.STOCK_MISSING_IMAGE , _('_Fechar sessão'), None,
                        _('Fechar sessão'), self.on_logout),
                ])


            self.uimanager.insert_action_group(actiongroup, 1)
            self.menu = self.uimanager.get_widget("/TrayIconMenu")
            self.reboot = self.uimanager.get_widget("/TrayIconMenu/reboot")
            pixbuf = self.icon_theme.load_icon('gnome-session-reboot', 0, 0)
            self.reboot.set_image(gtk.image_new_from_pixbuf(pixbuf))
            self.shutdown = self.uimanager.get_widget("/TrayIconMenu/shutdown")
            pixbuf = self.icon_theme.load_icon('gnome-session-halt', 0, 0)
            self.shutdown.set_image(gtk.image_new_from_pixbuf(pixbuf))
            self.logout = self.uimanager.get_widget("/TrayIconMenu/logout")
            pixbuf = self.icon_theme.load_icon('gnome-session-logout', 0, 0)
            self.logout.set_image(gtk.image_new_from_pixbuf(pixbuf))
            self.add(self.eventbox)

    def position_menu(self,menu):
        #Grab from deskbar applet
        align_to = self.eventbox
        direction = self.eventbox.get_direction()
        screen = menu.get_screen()
        monitor_num = screen.get_monitor_at_window(align_to.window)
        if monitor_num < 0:
                monitor_num = 0
        monitor = screen.get_monitor_geometry (monitor_num)
        menu.set_monitor (monitor_num)
        tx, ty = align_to.window.get_origin()
        twidth, theight = menu.get_child_requisition()
        tx += align_to.allocation.x
        ty += align_to.allocation.y
        if direction == gtk.TEXT_DIR_RTL:
                tx += align_to.allocation.width - twidth
```

Appendix D: Desktop Power Applet Code

```python
            if (ty + align_to.allocation.height + theight) <= monitor.y + monitor.height:
                ty += align_to.allocation.height
            elif (ty - theight) >= monitor.y:
                ty -= theight
            elif monitor.y + monitor.height - (ty + align_to.allocation.height) > ty:
                ty += align_to.allocation.height
            else:
                ty -= theight
        if tx < monitor.x:
            x = monitor.x
        elif tx > max(monitor.x, monitor.x + monitor.width - twidth):
            x = max(monitor.x, monitor.x + monitor.width - twidth)
        else:
            x = tx
        y = ty
        return (x, y, False)
    def on_button_press_event(self, item, event, data=None):
        self.menu.popup(None, None, self.position_menu, event.button, event.time)
    def on_lock(self, *args):
        os.system('xlock &')
    def on_shutdown(self, *args):
        dialog = gtk.MessageDialog(
                        parent=None,
                        flags=0,
                        type=gtk.MESSAGE_WARNING,
                        buttons=gtk.BUTTONS_YES_NO)
        dialog.set_resizable(False)
        dialog.set_title("")
        message_format = 'Você está prestes a desligar o computador'
        secondary_text = 'Você deseja continuar?'
        dialog.set_markup("<span weight='bold'size='larger'>%s</span>" \
                        % message_format)
        dialog.format_secondary_markup(secondary_text)
        result = dialog.run()
        if result == gtk.RESPONSE_YES:
            os.system('gdm-signal -h')
            os.system('pkill hildon-desktop')
        dialog.destroy()

    def on_logout(self, *args):
        dialog = gtk.MessageDialog(
                        parent=None,
                        flags=0,
                        type=gtk.MESSAGE_WARNING,
                        buttons=gtk.BUTTONS_YES_NO)
        dialog.set_resizable(False)
        dialog.set_title("")
        message_format = 'Você está prestes a deslogar do seu usuário'
        secondary_text = 'Você deseja continuar?'
        dialog.set_markup("<span weight='bold'size='larger'>%s</span>" \
                        % message_format)
        dialog.format_secondary_markup(secondary_text)
```

```python
            result = dialog.run()
            if result == gtk.RESPONSE_YES:
                os.system('pkill hildon-desktop')
            dialog.destroy()

    def on_reboot(self, *args):
        dialog = gtk.MessageDialog(
                        parent=None,
                        flags=0,
                        type=gtk.MESSAGE_WARNING,
                        buttons=gtk.BUTTONS_YES_NO)
            dialog.set_resizable(False)
            dialog.set_title("")
            message_format = 'Você está prestes a reiniciar o computador'
            secondary_text = 'Você deseja continuar?'
            dialog.set_markup("<span weight='bold'size='larger'>%s</span>" \
                        % message_format)
            dialog.format_secondary_markup(secondary_text)
            result = dialog.run()
            if result == gtk.RESPONSE_YES:
                os.system('gdm-signal -r')
                os.system('pkill hildon-desktop')
            dialog.destroy()


if __name__ == '__main__':
    icon = PowerTray()
    icon.show_all()
    try:
        gtk.main()
    except KeyboardInterrupt:
        print 'Exiting...'
        sys.exit(0)
```

D-Bus: An Overview

This is a condensed version of the canonical D-Bus tutorial found at http://dbus.freedesktop.org/doc/dbus-tutorial.html, which was written by Havoc Pennington, David Wheeler, John Palmieri and Colin Walters.

D-Bus is a system for *interprocess communication* (IPC). Architecturally, it has several layers:

- ❑ A library, libdbus, that allows two applications to connect to each other and exchange messages.

- ❑ A *message bus daemon* executable, built on libdbus that multiple applications can connect to. The daemon can route messages from one application to zero or more other applications.

- ❑ *Wrapper libraries* or *bindings* based on particular application frameworks — for example, libdbus-glib and libdbus-qt. There are also bindings to languages such as Python. These wrapper libraries are the API most people should use, as they simplify the details of D-Bus programming. libdbus is intended to be a low-level backend for the higher level bindings. Much of the libdbus API is useful only for binding implementation.

libdbus supports only one-to-one connections, just like a raw network socket. However, rather than sending byte streams over the connection, you send *messages*. Messages have a header identifying the kind of message, and a body containing a data payload. libdbus also abstracts the exact transport used (sockets versus whatever else), and handles details such as authentication.

The message bus daemon forms the hub of a wheel. Each spoke of the wheel is a one-to-one connection to an application using libdbus. An application sends a message to the bus daemon over its spoke, and the bus daemon forwards the message to other connected applications as appropriate. Think of the daemon as a router.

The bus daemon has multiple instances on a typical computer. The first instance is a machine-global singleton, that is, a system daemon similar to sendmail or Apache. This instance has heavy security restrictions on what messages it will accept, and is used for system-wide communication.

Appendix E: D-Bus: An Overview

The other instances are created one per user login session. These instances allow applications in the user's session to communicate with one another.

The system-wide and per-user daemons are separate. Normal within-session IPC does not involve the system-wide message bus process and vice versa.

D-Bus Applications

D-Bus is designed for two specific cases:

- Communication between desktop applications in the same desktop session; to allow integration of the desktop session as a whole, and address issues of process lifecycle (When do desktop components start and stop running?)
- Communication between the desktop session and the operating system, where the operating system would typically include the kernel and any system daemons or processes

For the within-desktop-session use case, the GNOME and KDE desktops have significant previous experience with different IPC solutions such as CORBA and DCOP. D-Bus is built on that experience and tailored to meet the needs of these desktop projects.

The problem solved by the system-wide or communication-with-the-OS case is best explained by the following text from the Linux Hotplug project:

> A gap in current Linux support is that policies with any sort of dynamic "interact with user" component aren't currently supported. For example, that's often needed the first time a network adapter or printer is connected, and to determine appropriate places to mount disk drives. It would seem that such actions could be supported for any case where a responsible human can be identified: single user workstations, or any system which is remotely administered.

This is a classic "remote sysadmin" problem, where in this case hotplugging needs to deliver an event from one security domain (operating system kernel, in this case) to another (desktop for a logged-in user, or remote sysadmin). Any effective response must go the other way: the remote domain taking some action that lets the kernel expose the desired device capabilities. (The action can often be taken asynchronously — for example, letting new hardware be idle until a meeting finishes.) At this writing, Linux doesn't have widely adopted solutions to such problems. However, the new D-Bus work may begin to solve that problem.

D-Bus might be useful for purposes other than the one for which it was designed (see Figure E-1). General properties that distinguish it from other forms of IPC are:

- Binary protocol designed to be used asynchronously (similar in spirit to the X Window System protocol).
- Stateful, reliable connections held open over time.
- The message bus is a daemon, not a "swarm" or distributed architecture.
- Many implementation and deployment issues are specified rather than left ambiguous/configurable/pluggable.

Appendix E: D-Bus: An Overview

- Semantics are similar to the existing DCOP system, allowing KDE to adopt it more easily.
- Security features to support the system-wide mode of the message bus.

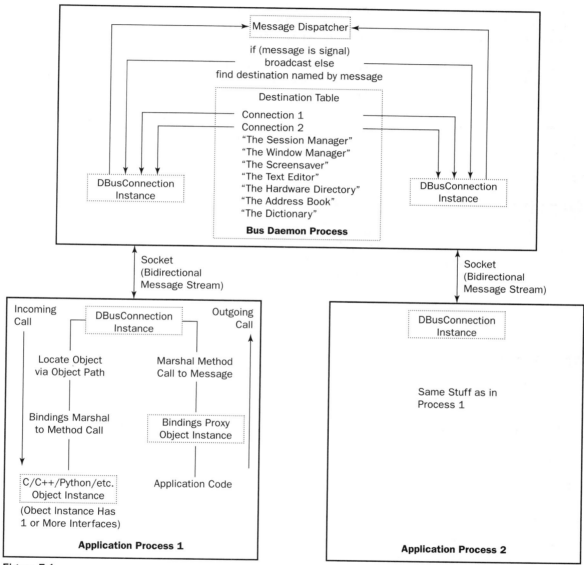

Figure E-1

Appendix E: D-Bus: An Overview

Native Objects and Object Paths

Your programming framework defines what an "object" is like, usually with a base class — for example: java.lang.Object, GObject, QObject, Python's base Object, or whatever. Let's call this a *native object*.

The low-level D-Bus protocol, and corresponding libdbus API, does not care about native objects. However, it provides a concept called an *object path*. The idea of an object path is that higher-level bindings can name native object instances, and allow remote applications to refer to them.

The object path looks like a filesystem path — for example, an object could be named /org/kde/kspread/sheets/3/cells/4/5. Human-readable paths are nice, but you are free to create an object named /com/ubuntu/c5yo817y0c1y1c5b if it makes sense for your application.

Namespacing object paths by starting them with the components of a domain name you own (e.g., /org/kde) is nice as it keeps different code modules in the same process from stepping on one another's toes.

Methods and Signals

Each object has *members*; the two kinds of members are *methods* and *signals*. Methods are operations that can be invoked on an object, with optional input (aka arguments or "in parameters") and output (aka return values or "out parameters"). Signals are broadcast from the object to any interested observers of the object; signals may contain a data payload.

Both methods and signals are referred to by name, such as PUMD or OnClicked.

Interfaces

Each object supports one or more *interfaces*. Think of an interface as a named group of methods and signals, just as it is in GLib or Qt or Java. Interfaces define the *type* of an object instance.

D-Bus identifies interfaces with a simple namespaced string, something like org.freedesktop.Introspectable. Most bindings will map these interface names directly to the appropriate programming language construct — for example, to Java interfaces or C++ pure virtual classes.

Proxies

A *proxy object* is a convenient native object created to represent a remote object in another process. The low-level D-Bus API involves manually creating a method call message, sending it, and then manually receiving and processing the method reply message. Higher-level bindings provide proxies as an alternative. Proxies look like a normal native object, but when you invoke a method on the proxy object, the binding converts it into a D-Bus method call message, waits for the reply message, unpacks the return value, and returns it from the native method.

In pseudocode, programming *without* proxies might look like this:

```
Message message = new Message("/remote/object/path", "MethodName", arg1, arg2);
Connection connection = getBusConnection();
connection.send(message);
Message reply = connection.waitForReply(message);
if (reply.isError()) {

} else {
   Object returnValue = reply.getReturnValue();
}
```

However, programming *with* proxies might look like this:

```
Proxy proxy = new Proxy(getBusConnection(), "/remote/object/path");
Object returnValue = proxy.MethodName(arg1, arg2);
```

Bus Names

When each application connects to the bus daemon, the daemon immediately assigns it a name, called the *unique connection name*. A unique name begins with a colon (:) character. These names are never reused during the lifetime of the bus daemon — that is, you know a given name will always refer to the same application. An example of a unique name might be :34-907. The numbers after the colon have no meaning other than their uniqueness.

When a name is mapped to a particular application connection, that application is said to *own* that name.

Applications may ask to own additional *well-known names*. For example, you could write a specification to define a name called com.ubuntu.TextEditor. Your definition could specify that to own this name, an application should have an object at the path /com/ubuntu/TextFileManager supporting the interface org.freedesktop.FileHandler.

Applications could then send messages to this bus name, object, and interface to execute method calls.

You could think of the unique names as IP addresses, and the well-known names as domain names. So com.ubuntu.TextEditor might map to something like :34-907 just as ubuntu.com maps to something like 91.189.94.250.

Names have a second important use, other than routing messages. They are used to track lifecycle. When an application exits (or crashes), its connection to the message bus will be closed by the operating system kernel. The message bus then sends out notification messages telling the remaining applications that the application's names have lost their owner. By tracking these notifications, your application can reliably monitor the lifetime of other applications.

Bus names can also be used to coordinate single-instance applications. If you want to be sure only one com.mycompany.TextEditor application is running, for example, have the text editor application exit if the bus name already has an owner.

Appendix E: D-Bus: An Overview

Addresses

Applications using D-Bus are either servers or clients. A server listens for incoming connections; a client connects to a server. Once the connection is established, it is a symmetric flow of messages; the client-server distinction only matters when setting up the connection.

If you're using the bus daemon, as you probably are, your application will be a client of the bus daemon. That is, the bus daemon listens for connections and your application initiates a connection to the bus daemon.

A D-Bus *address* specifies where a server will listen, and where a client will connect. For example, the address `unix:path=/tmp/abcdef` specifies that the server will listen on a UNIX domain socket at the path /tmp/abcdef and the client will connect to that socket. An address can also specify TCP/IP sockets, or any other transport defined in future iterations of the D-Bus specification.

When using D-Bus with a message bus daemon, `libdbus` automatically discovers the address of the per-session bus daemon by reading an environment variable. It discovers the system-wide bus daemon by checking a well-known UNIX domain socket path (although you can override this address with an environment variable).

If you're using D-Bus without a bus daemon, it's up to you to define which application will be the server and which will be the client, and specify a mechanism for them to agree on the server's address. This is an unusual case.

Big Conceptual Picture

To specify a particular method call on a particular object instance, a number of nested components have to be named:

```
Address -> [Bus Name] -> Path -> Interface -> Method
```

The bus name is in brackets to indicate that it's optional — you only provide a name to route the method call to the right application when using the bus daemon. If you have a direct connection to another application, bus names aren't used; there's no bus daemon.

The interface is also optional, primarily for historical reasons; DCOP does not require that you specify the interface; instead, it simply forbids duplicate method names on the same object instance. D-Bus will thus let you omit the interface, but if your method name is ambiguous, it is undefined which method will be invoked.

Messages: Behind the Scenes

D-Bus works by sending messages between processes. If you're using a sufficiently high-level binding, you may never work with messages directly.

Appendix E: D-Bus: An Overview

There are four message types:

- Method call messages ask to invoke a method on an object.
- Method return messages return the results of invoking a method.
- Error messages return an exception caused by invoking a method.
- Signal messages are notifications that a given signal has been emitted (that an event has occurred). You could also think of these as "event" messages.

A method call maps very simply to messages: You send a method call message, and receive either a method return message or an error message in reply.

Each message has a *header*, including *fields*, and a *body*, including *arguments*. You can think of the header as the routing information for the message, and the body as the payload. Header fields might include the sender bus name, destination bus name, method or signal name, and so forth. One of the header fields is a *type signature* describing the values found in the body. For example, the letter "i" means "32-bit integer" so the signature "ii" means the payload has two 32-bit integers.

Calling a Method: Behind the Scenes

A method call in D-Bus consists of two messages: a method call message sent from process A to process B, and a matching method reply message sent from process B to process A. Both the call and the reply messages are routed through the bus daemon. The caller includes a different serial number in each call message, and the reply message includes this number to allow the caller to match replies to calls.

The call message will contain any arguments to the method. The reply message may indicate an error, or may contain data returned by the method.

The steps of a method invocation in D-Bus are as follows:

1. The language binding may provide a proxy, such that invoking a method on an in-process object invokes a method on a remote object in another process. If so, the application calls a method on the proxy, and the proxy constructs a method call message to send to the remote process.
2. For more low-level APIs, the application may construct a method call message itself, without using a proxy.
3. In either case, the method call message contains: a bus name belonging to the remote process; the name of the method; the arguments to the method; an object path inside the remote process; and, optionally, the name of the interface that specifies the method.
4. The method call message is sent to the bus daemon.
5. The bus daemon looks at the destination bus name. If a process owns that name, the bus daemon forwards the method call to that process. Otherwise, the bus daemon creates an error message and sends it back as the reply to the method call message.
6. The receiving process unpacks the method call message. In a simple low-level API situation, it may immediately run the method and send a method reply message to the bus daemon. When using a high-level binding API, the binding might examine the object path, interface, and

Appendix E: D-Bus: An Overview

method name, and convert the method call message into an invocation of a method on a native object (`GObject`, `java.lang.Object`, `QObject`, and so on), and then convert the return value from the native method into a method reply message.

7. The bus daemon receives the method reply message and sends it to the process that made the method call.

8. The process that made the method call looks at the method reply and makes use of any return values included in the reply. The reply may also indicate that an error occurred. When using a binding, the method reply message may be converted into the return value of a proxy method, or into an exception.

9. The bus daemon never reorders messages. That is, if you send two method call messages to the same recipient, they will be received in the order they were sent. The recipient is not required to reply to the calls in order, however; for example, it may process each method call in a separate thread, and return reply messages in an undefined order depending on when the threads complete. Method calls have a unique serial number used by the method caller to match reply messages to call messages.

Emitting a Signal: Behind the Scenes

A signal in D-Bus consists of a single message, sent by one process to any number of other processes. That is, a signal is a unidirectional broadcast. The signal may contain arguments (a data payload), but because it is a broadcast, it never has a "return value." Contrast this with a method call where the method call message has a matching method reply message.

The emitter (aka sender) of a signal has no knowledge of the signal recipients. Recipients register with the bus daemon to receive signals based on "match rules" — these rules would typically include the sender and the signal name. The bus daemon sends each signal only to recipients who have expressed interest in that signal.

The process of emitting a signal in D-Bus occurs in the following manner:

1. A signal message is created and sent to the bus daemon. When using the low-level API this may be done manually; with certain bindings it may be done for you by the binding when a native object emits a native signal or event.

2. The signal message contains the name of the interface that specifies the signal, the name of the signal, the bus name of the process sending the signal, and any arguments.

3. Any process on the message bus can register "match rules" indicating which signals it is interested in. The bus has a list of registered match rules.

4. The bus daemon examines the signal and determines which processes are interested in it. It sends the signal message to these processes.

Each process receiving the signal decides what to do with it; if using a binding, the binding may choose to emit a native signal on a proxy object. If using the low-level API, the process may just look at the signal sender and name, and decide what to do based on what it sees.

Introspection

D-Bus objects may support the interface org.freedesktop.DBus.Introspectable. This interface has one method, `Introspect`, which takes no arguments and returns an XML string. The XML string describes the interfaces, methods, and signals of the object.

Python and D-Bus

There are two bus daemons of interest. Each user login session should have a *session bus*, which is local to that session. It is used to communicate between desktop applications. Connect to the session bus by creating a `SessionBus` object:

```
import dbus

session_bus = dbus.SessionBus()
```

The *system bus* is global and usually started during boot; it's used to communicate with system services such as udev, Network Manager, and HAL. To connect to the system bus, create a `SystemBus` object:

```
import dbus

system_bus = dbus.SystemBus()
```

Of course, you can connect to both in the same application.

Method Calls

D-Bus applications can export objects for other applications' use. To start working with an object in another application, you need to know the following:

- **The bus name** — This identifies the application that you want to communicate with. You'll usually identify applications by a well-known name, which is a dot-separated string starting with a reversed domain name, such as org.freedesktop.NetworkManager or com.example.WordProcessor.
- **The object path** — Applications can export many objects. For instance, example.com's word processor might provide an object representing the word processor application itself and also an object for each document window that is opened, or it might also provide an object for each paragraph that's within a document.

To identify which one you want to interact with, you use an object path, a slash-separated string resembling a filename. For instance, example.com's word processor might provide an object at / representing the word processor itself, and objects at /documents/123 and /documents/345 representing opened document windows.

As you'd expect, one of the main things you can do with remote objects is to call their methods. As in Python, methods may have parameters, and they may return one or more values.

Appendix E: D-Bus: An Overview

Proxy Objects

To interact with a remote object, you use a *proxy object*. This is a Python object that acts as a proxy or "stand-in" for the remote object. When you call a method on a proxy object, this causes dbus-Python to make a method call on the remote object, passing back any return values from the remote object's method as the return values of the proxy method call.

To obtain a proxy object, call the `get_object` method on the `Bus`. For example, NetworkManager has the well-known name org.freedesktop.NetworkManager and exports an object whose object path is /org/freedesktop/NetworkManager, plus an object per network interface at object paths such as /org/freedesktop/NetworkManager/Devices/eth0. You can get a proxy for the object representing eth0 like this:

```
import dbus
bus = dbus.SystemBus()
proxy = bus.get_object('org.freedesktop.NetworkManager',
                       '/org/freedesktop/NetworkManager/Devices/eth0')
# proxy is a dbus.proxies.ProxyObject
```

Interfaces and Methods

D-Bus uses *interfaces* to provide a namespacing mechanism for methods. An interface is a group of related methods and signals (more on signals later), identified by a name that is a series of dot-separated components starting with a reversed domain name. For instance, each NetworkManager object representing a network interface implements the interface org.freedesktop.NetworkManager.Devices, which has methods such as `getProperties`.

To call a method, call the method of the same name on the proxy object, passing in the interface name via the `dbus_interface` keyword argument:

```
import dbus
bus = dbus.SystemBus()
eth0 = bus.get_object('org.freedesktop.NetworkManager',
                      '/org/freedesktop/NetworkManager/Devices/eth0')
props = eth0.getProperties(dbus_interface='org.freedesktop.NetworkManager.Devices')
# props is a tuple of properties, the first of which is the object path
```

As a short cut, if you're going to be calling many methods with the same interface, you can construct a `dbus.Interface` object and call methods on that, without needing to specify the interface again:

```
import dbus
bus = dbus.SystemBus()
eth0 = bus.get_object('org.freedesktop.NetworkManager',
                      '/org/freedesktop/NetworkManager/Devices/eth0')
eth0_dev_iface = dbus.Interface(eth0,
    dbus_interface='org.freedesktop.NetworkManager.Devices')
props = eth0_dev_iface.getProperties()
# props is the same as before
```

Index

Index

! (exclamation marks), seed files and, 232
* (asterisks), packages in raw seed lists and, 231

A

Abreau, John, 271
Acer Aspire One test results, 162–163
actions, configuring (pbuilder), 125
addresses (D-Bus), 310
administration, repository (Git), 285
Adobe Flash, 213–214
Alternate Image, 226
animation
    Clutter and, 70–74
    gadget with, 136–137
Apple, history of
    history of, 3–4
applet for mounting removable media (example), 94–103
application autostart specification, desktop, 57
application development
    applet example, 94–103
    applications, creating. See applications, creating
    D-Bus. See D-Bus
    GConf, 89–91
    notifications, 93–94
    overview, 53
    Ubuntu mobile releases, 54–55
    Ubuntu Netbook Remix, 54
application packaging
    pbuilder. See pbuilder tool
    Personal Package Archives (PPA) and, 108
    PPA. See PPA (Personal Package Archives)
    repositories, creating, 118–119
    tools for, 105–108
application selection
    business users, 129–131
    location-aware users, 138. See also GPS-enabled standalone web application (MID devices)
    multimedia users, 131–133
    social network users, 133–137
applications
    D-Bus, 306–307
    installing inside of images, 33
    troubleshooting icon not appearing, 258–259
applications, creating
    design, 55–56
    free desktop standards, 56–58
    GTK. See GTK (Gimp Toolkit)
    Hildon Application Framework, 58–60
Apport's application, 206
approx tool, 225–226
apt (Advanced Package Tool), 106
    installing packages on, 261–262
architecture
    ARM architectures, 250–251
    checking support of, 220
    defined, 243
    fine-tuning the kernel and, 221
    mobile computing platform, 243–244
archive of packages (Ubuntu), 126
Arima, running Ubuntu on, 261
ARM (Advanced RISC Machines) platform
    ARM architectures, 250–251
    installing on QEMU, 22–23
    overview, xxv
    SheevaPlug and, 265–276
Assistive Technology Service Provider Interface (AT-SPI), 188

atime, disabling, 212
ATK. See LSB Application Testkit (ATK)
AT-SPI (Assistive Technology Service Provider Interface), 188
automatic repositories, creating, 118
automatic setting of flags, 259–260
automatic theming (MID), 156–157
autostart specification (desktop application), 57
awake state (power), 36

B

background services, turning off, 212–213
backporting KVM, 119
Barton, George, 254
batteries
    basics, 210
    comparing, 48–50
    recharging, 209
    testing, 47–50
Bazaar
    Launchpad and, 288–289
    to obtain scripts, 30–31
bind mounting directories (pbuilder), 126
bindings (D-Bus), 305
blacklist files, 233–234
blob object (Git), 277
Bluetooth
    GPS application and, 138, 142
    turning on/off, 45–46
Bogstad, Bill, 272
booting
    boot selector, 222–224
    boot speed, 207
    dual boots, 259
    MID devices and theming, 149
    troubleshooting, 257–258
branch command (Git), 282–283
browsers, testing with Mago, 193–195
bugs, reporting, 205–206
build process, fine-tuning, 225–226
build-essential package (dpkg), 107
Burton, Ross, 142
bus names (D-Bus), 309

business
    MIDs as opportunity, 129
    mobile market unpredictability, 243
    users of MIDs and, 129–131

C

caching packages, 225–226
callback functions, signals and, 66
Canola (application), 81, 131
cellphones. See also iPhones
    ARM architectures and, 250
    history of, 4
    vs. PCs, 255
certification, LSB and, 196–197
changelog file, 112, 238
changeset object, (Git), 278
checkout command (Git), 283
Clearlooks, theming and, 157
cloning existing repositories (Git), 279
Cloud computing, 250
Clutter library, xxv, 70–75
colors, theming and, 153, 155
command line, reporting bugs from, 205–206
commands
    apt, listed, 106
    DKMS, 183–185
    dpkg, listed, 106–107
compilation, defined, 241
compliance, 204–205
./configure application, 105
configuring
    actions (pbuilder), 125
    default configuration for Ubuntu/xfce/Hildon behaviors, 224
    GConf (default), 224
    pbuilder, 120–123
    pre-configuring GDM, 224
    repository configuration (Git), 285
    touchscreens, 214–217
Connecting to the Net.Generation: What higher education professionals need to know about today's students, 134
contributor, commands for (Git), 284

control file (PPA), 112–113
copying gadgets, 136
copyright
  copyright file (PPA), 113–114
  Ubuntu, 240–242
CPUFREQ and governors, 211
CPUs
  CPUFREQ and, 211
  Flash and, 213–214
cross compiler, defined, 22
custom distribution of Ubuntu Mobile. *See* Ubuntu Mobile image example

D

Darwin, Charles, 247
Darwin Model of Software Development, 247
D-Bus
  addresses, 310
  applications, 306–307
  basics, 305–306
  bus names, 83–84, 309
  D-Bus Send, 87
  D-Bus Viewer, 85–86
  dbus-launch command, 88
  dbus-monitor command, 87
  D-feet debugger, 88
  exporting objects with, 84–85
  GPS application and, 138, 143–145
  interfaces, 308
  interfaces and, 314
  Introspect method, 313
  messages and, 310–311
  method calls in, 311–312, 313
  methods and signals, 308
  native objects/object paths, 308
  nested components, naming, 310
  object paths, 83–84
  overview, 82
  proxy objects, 308–309, 314
  Python and, 313
  security, 89–91
  signals, connecting to, 85
  signals in, 312

dch tool, 238
dd for Windows, 260
debhelper suite of tools, 112
Debian
  debian-installers, 226, 239
  kernel packages, creating, 172–175
  repositories, setting up, 225
Debian files, 114
debugging. *See also* **troubleshooting**
  D-feet debugger, 88
derived distribution, 240–241
design, application, 55–56
desktop
  application autostart specification, 57
  entry specification, 56–57
  free desktop standards, 56–58
  menu specification, 57–58
  XDG base directory specification, 57
.desktop files in Python, 74
Desktop Image, 226
desktop power applet code (Python), 299–303
DeviceKit-power, 41–44
devscripts tool, 107
D-feet debugger, 88
dh_ commands, 111–112
d-i (debian-installer), 226
Diamond, David, 246
diff program (dpkg), 107
directories, XDG base specification, 57
Display Manager, 223–224
distribution, derived, 240–241
distribution environments, creating (pbuilder), 124
Django web application framework, 138, 139, 143–144
DKMS (dynamic kernel module support). *See* **dynamic kernel module support (DKMS)**
DKMS (Dynamic Kernel Modules Support). *See* **dynamic kernel module support (DKMS)**
DNS and caching server, setting up, 27
documents, OpenOffice and, 130
downloading kernel source, 166

dpkg tools
   for downloading kernel source, 175
   dpkg-scanpackages, 107
   dpkg-scansources, 107
   for packaging, 106–107
dput tool, 115
drivers, updating (DKMS), 181–185
dual boots, 259
dynamic kernel module support (DKMS)
   commands, 183–185
   framework, internal workings of, 182–183
   overview, 181–182

E

Eee PC test results, 162–163
EFL (Enlightenment Foundation Libraries), 79–81
Elementary widget set (Enlightenment suite), 81–82
embedded systems, 221–222
encryption, setting up, 130
Enderle, Rob, 266
energy tips, 208–212
engines, theme, 148, 157
Enlightenment Foundation Libraries. See EFL (Enlightenment Foundation Libraries)
Entertainer media center example, 131–133
entry specification, desktop, 56–57
event loops (GTK), 64–65
exporting objects with D-Bus, 84–85

F

fakeroot tool, 108
Feldman, Jerry, 271
file systems
   embedded, 221–222
   location of themes, 148
files
   office files. See treb (Trebuchet) application
   raw seed files, 231–232
   seed files, 231
   sharing between guests and hosts, 29–31

fixing problems. See problems and solutions
flags (USE_HILDON), setting automatically, 259–260
Flash, Adobe, 213–214
Frankenstein Model of software development, 248
free (memory) application, 202
free desktop standards, 56–58
free software, 247, 251
Freerunner, running Ubuntu on, 260
fsck command (Git), 279–280
future
   of Linux, 245, 253, 255
   of mobile industry, 253–254

G

gadgets
   Google Gadgets, 134, 135–137
   gOS 3 Gadgets, 134–138
gc command (Git), 280
GConf
   booting MID devices and, 149
   default configuration, setting, 224
   overview, 91–94
   user interface and, 54–55
GDM display manager
   pre-configuring, 224
   themes, creating, 158–159
Generation Y (Gen Y), 134
geohash.org, 145
Germinate program, 231–235
germinate tool, 234, 235
Git
   contributor commands, 284
   individual developer commands, 280–284
   integrator role, 284
   maintainer role, 284–285
   objects, 277–278
   overview, 277
   repository administration, 285
   repository commands, 278–280
   repository configuration, 285
   tool, 171–175
   tree, 173

Glade, 190–191
  .glade file (pumdGlade class), 69–70
  Interface Designer, 66–70
GNOME. See also **GConf; Glade**
  Gnome-Power-Manager, 40
  Gnome-Power-Statistics application, 40–41
  Hildon Application Framework and, 57
GNU Haret, 259
gnupg tool, 107
Google Gadgets, 134, 135–137
gOS 3 Gadgets, 134–137
governors, CPUFREQ and, 211
gpsd daemon, 138
GPS-enabled standalone web application (MID devices), 138–145
  background, 138–139
  D-Bus and HTTP requests, 143–145
  implementation, 139–141
  interaction with GPS daemon, 142–143
  testing gypsy to GPS connection, 142
graphical corruption, preventing, 262
graphics, Vesa drivers, 262
Grub2 boot selector, 222–223
GTK (Gimp Toolkit)
  Glade Interface Designer and, 66–70
  GTK engines, theming and, 148
  GTK+, 64, 147
  gtkrc file, customizing, 152–153
  GtkWidget, theming and, 154
  horizontal boxes, 65–66
  layout and, 65
  overview, 64–65
  signals, 65
  vertical/horizontal boxes, 66
gypsy daemon
  basics, 139–140
  testing connection to GPS, 142

H

hard coding modules, 207–208
hard disks
  adding in VirtualBox, 16–17
  watching activity of, 217–218
hardware
  architectures, 220
  checking support of, 220
  dual boots and, 259
  hardware abstraction layer (hal), defined, 215
  Hardware Compatibility Lists (HCL), 220
hdparm, 211
Hello World application (Hildon), 60–64
hibernate power mode, 36–40
hibernation power mode (Hildon), 60
Hildon Application Framework
  overview, xxiii, 58
  themes and, 155, 156–157
  user interface, default behaviors, 224
Holmes, Iain, 139
hook commands/scripts, 125–126
horizontal boxes, GTK and, 66
hosting projects on Launchpad, 287–289
HTTP requests, GPS application and, 143–145
Human metacity theme, 160–163
Human Netbook Theme, 54
hwinfo program, 220
hybrid-suspend power mode, 37–38
Hypervisor software, 11

I

icons, application, 258–259
images. See also **multimedia users**
  building customized, 32–33, 230–235
  creating, 226–230
  default Ubuntu image, creating, 226–230
  .img images, 231
  increasing downloaded size of, 34
  installing applications inside of, 33
  modifying when theming, 155–156
  scripts for working with, 31–32
  theme images, sapwood and, 148–149
  writing to USB sticks, 257–258
Independent software vendor (ISV), defined, 187
init command (Git), 279
initramfs, 207
installers
  pre-seeding, 239–240
  selecting, 226

installing
  applications inside of images, 33
  ARM on QEMU, 22–23
  KVM, 23–24
  LSB Application Testkit (ATK), 196
  Ubuntu MID image, 24
  Ubuntu Netbook Remix on UMPCs, 261
  VirtualBox, 12
integrator role (Git), 284
interfaces
  changes to GUI, 78
  D-Bus, 308, 314
Internet Relay Chat (IRC), defined, 288
interprocess communication (IPC), 305
Intrepid, xxv
Introspect method (D-Bus), 313
invention
  Invention: The Care and Feeding of Ideas, 244
  stages of, 244–245
  Torvalds and, 246
iotop application, 217
iPhones
  on campus, 253, 254
  history of, 3–4, 4
iPods, history of, 3–4
ISO images
  building custom, 240
  building default, 229–231
ISV, defined, 187

J

Jaunty Ubuntu Mobile
  defined, xxv
  MID release, downloading, 12
Junco, Reynol, 134
Just for Fun: The Story of an Accidental Revolutionary, 246

K

Karmic, Ubuntu. See Ubuntu Mobile image example
kernel fine-tuning
  Debian package, creating, 172–175

  dynamic kernel module support. *See* dynamic kernel module support (DKMS)
  kernels defined, 165
  Linux kernels, 165
  non-Ubuntu kernel tree, updating, 181
  overview, 165–166
  reasons for, 165
  Ubuntu Karmic example and, 221
  Ubuntu kernel tree, updating, 175–181
  Ubuntu package, creating, 166–171
keyboards
  Entertainer controls on, 133
  onscreen, enabling, 224
Khoury, Rabeeh, 269
Kidder, Tracy, 246
KVM (Kernel Virtual Machine)
  backporting, 119
  basics, 21–24
  KVM/QEMU, networking in, 26–34
  troubleshooting, 262
  using bridge in, 29

L

laptop mode, 209–210
Latencytop tool, 203
launcher. *See* **Ubuntu Netbook Remix**
Launchpad. *See also* **PPA (Personal Package Archives)**
  hosting projects on, 287–289
layouts, GTK+ and, 65
LDTP (Linux Desktop Testing Project), 187
LGPL (Library General Public License), 241–242
libraries, libdbus (D-Bus), 305
lintian tool, 108
Linux. *See also* **Moblin initiative**
  future of, 245, 253, 255
  Hotplug project, 306
  Linux Desktop Testing Project (LDTP), 187
  LSB Application Testkit (ATK), 196–197
  mobile computing and, 254–255
  performance testing tools, 201–203
  story of, 246
  student use of, 254

Linux Standard Base (LSB) Testing Toolkit, 187
location-aware users, 138. *See also* GPS-enabled standalone web application (MID devices)
Long Term Support (LTS) releases, 219
Lotus Development, 247
LPIA (Low Power Intel Architecture)
  flags, setting automatically, 259
  LPIA Ubuntu MID release, 50
  overview, xxiii
  Ubuntu MID kernel and, 165
LSB Application Testkit (ATK), 196–197
lshw program, 220
luvcview application, 220

M

Mago, testing with, 188–189, 193–195
main package category (Ubuntu), 126
maintainer role (Git), 284–285
make tool, 105–106
manual theming of MID, 154–156
marketplace (mobile devices)
  growth in popularity of, 243
  razors and blades pricing, 251
  unpredictability of, 243
Mastrodicasa, Jeanna, 134
Maximus daemon, 54
Mayr, Ernst, 248
Mayr Model of software development, 248
McCaslin platform, xxiv
memcheck, 202
memstat application, 202
Mendel, Gregor Johann, 247
Mendel Model of software development, 247–248
Menlow platform, xxiv
menu specification, desktop, 57–58
menus, creating externally (Hildon), 59
menus, creating (Hello World), 61–62
messages
  message bus daemons (D-Bus), 305
  sending (D-Bus), 310–311
metacity-theme-viewer, 160–163
metapackages
  building, 236–238
  generating, 235–236, 238–239
methods
  interfaces and (D-Bus), 314
  method calls (D-Bus), 311–313
  and signals (D-Bus), 308
MID, Ubuntu, 54–55
mobile computing
  architecture, 243–244
  devices, 253–255
mobile technology
  history of, 2–5
  market unpredictability, 243
  Mobile Developers team, joining, 262
  mobile environment. *See* Ubuntu mobile environment
  Mobile Internet Devices (MIDs), 1, 129
  Obama and, 249
  overview, 1
  politics of, 249
  Shuttleworth on, 254–255
Moblin initiative, xxiii, 250–251, 255,
modules. *See also* dynamic kernel module support (DKMS)
  hard coding, 207–208
MONITOR environmental variable, 48
mount binding package repositories (pbuilder), 126
"The Moving Target Problem", xxi
Mozilla Prism, 135
multimedia users, 131–133
multiverse package category (Ubuntu), 127
Murrine engine, theming and, 157

N

naming
  bus names, 83
  bus names (D-Bus), 309
native objects/object paths (D-Bus), 308
Neo Freerunner, 260
netbooks
  history of, 6
  Linux, Ubuntu and, 6–8

networking

networking
    connections, 251–252
    in KVM/QEMU, 26–34
    in VirtualBox, 25–34
The Next Billion Network, 249
Niemeyer, Gustavo, 145
Nokia Wireless GPS Module LD-3W, 142
notifications (D-Bus), 93–94

O

objects
    exporting with D-Bus, 84–85
    Git, 277–278
    object paths, D-Bus, 83
On the Origin of Species, 247
Open Source Software (OSS), 249, 251–252
OpenCV (Open Source Computer Vision), 267–268
OpenOffice, 130–131

P

P2P forums, xxvii
packages
    adding to/removing from repositories, 119
    building (PPA), 115
    building inside chroot (pbuilder), 124
    caching with approx, 225–226
    categories of, 126–127
    germinating seeds and, 234
    installing on apt, 261–262
    metapackages, building, 236–238
    metapackages, generating, 235–236, 238–239
    Packages.gz, 118
    pbuilder tool for testing, 200
    power management, 36–41
    sections of (Ubuntu), 127
packaging applications. *See* application packaging
padding, defined, 153
Palmieri, John, 305
passwords, 130
patch program (dpkg), 107
Paul, Ryan, 157

PBuilder
    hook manipulation with, 125–126
    mount binding package repositories for use with, 126
pbuilder tool
    configuring, 120–123
    overview, 120
    as package builder, 108
    performing actions on, 123–126
    for testing packages, 200
pdebuild, 124–125
Pennington, Havoc, 305
performance testing themes, 160–163
Phoronix Test Suite, 47–50, 197–200
plug computing, 265–266
pmi (powermanagement-interface), 38
pm-utils power package, 37–40
policies
    Ubuntu, 240–242
    Ubuntu on packaging, 126–127
PolicyKit (D-Bus), 90–91
politics of technology, 249
Port to arm (Ubuntu), 54
power management. *See also* batteries
    DeviceKit-power, 41–44
    on disks, 211–212
    Gnome-Power-Manager, 40
    Gnome-Power-Statistics application, 40–41
    investigating power usage, 46–47
    overview, 35–36
    packages, 36–41
    pmi, 38
    policies, defining, 221
    Power Save Poll protocol (PS-Poll), 44
    power saving states, 36
    powertop, 46–47
    radio transmitters, controlling, 44–46
PPA (Personal Package Archives)
    changelog file, 112
    control file, 112–113
    copyright file, 113
    defined, xxiv
    overview, 108
    packages, building, 115
    REVU tool, 116–117

RFA packages, 117
rules, 108–112
uploading to, 115–116
pre-seeding installers, 239–240
Primary Master hard disk (VirtualBox), 15
problems and solutions
application icon not appearing, 258–259
boot process stopping, 257–258
graphical corruption, 262
installing packages on apt, 261–262
KVM, 262
poor performance, 262
QEMU, 262
Procinfo tool, 201
program-wide settings (Hildon), 59–60
proxy objects (D-Bus), 308–309, 314
prune command (Git), 280
ps tool, 201
pumdGlade class, 67–69
Python
D-Bus and, 313
packages, installing, 188

Q

QEMU (QuickEmulator)
basics, 21–24
installing, xxvi
installing ARM on, 22–23
networking in, 26–34
troubleshooting, 262
QoS interface (DeviceKit-power), 43–44
QT overview, 75–79

R

Radachowsky, Sage, 272
radio transmitters, controlling, 44–46
RAM, /tmp and, 208
raw seed files, 231–233
Raymond, Eric, 248
razors and blades markets, 251
recharging devices, 209
releases, Ubuntu, 219
Relocatable Computing, 268
reporting bugs, 205–206

repositories
creating automatic, 118
creating local, 118
repository administration (Git), 285
repository commands (Git), 278–280
repository configuration (Git), 285
setting up Debian, 225
reprepro tool, 118
restricted package category (Ubuntu), 126
REVU tool, 116–117
RFA (Request For Adoption) packages, 117
RFKILL (WiFi), 45
rules file (PPA), 109–112

S

sapwood engine, 148
SCM (Source Code Management), 277
screen size in application design, 55
scripts for images, 30–31
sections of package categories (Ubuntu), 127
security
business users and, 130
D-Bus, 89–91
seed germination, 231–235
Sensory Overload, 249
services (background), turning off, 212–213
session bus (D-Bus), 313
SessionBus Bus type, 82
SheevaPlug
application ideas for, 267–268
background, 252
background of, 265–266
components of initial kit, 265–266
Relocatable Computing, 268
solar-powered cluster, building, 270–276
SWARM, 268–276
Shelly, Mary, 248
show-branch command (Git), 281
Shuttleworth, Mark, 8, 75, 248, 254–255
signals (D-Bus), 308, 312
signals, connecting to D-Bus, 85
social network users, 133–137
software
evolution and development of, 246–248
open source, 251

solar-powered cluster, building (SWARM), 270–276
"Soul of a New Machine" (Kidder), 246
source code, where to find, xxvi
squashfs filesystem, 32
Stallman, Richard, 247
standby state (power), 36
states, power saving, 36
STRUCTURE files, 232–234, 237
styles, theming and, 153–154
suspend state (power), 36–40
SWARM, 268–276
SystemBus Bus type, 82, 313
systems, embedded, 221–222

T

technology world
  changes in, 247
  Obama and, 249
  politics and, 249
templates for copyright files, 114
testing
  accessibility libraries for, 188
  application for, building. See testing application, building (example)
  batteries, 47–50
  bug reporting, 205–206
  compliance, 204–205
  free application for, 202
  gypsy connection to GPS, 142
  Latencytop tool, 203
  memcheck, 202
  memstat application, 202
  pbuilder for package testing, 200
  Phoronix Test Suite, 197–200
  Procinfo tool, 201
  ps tool, 201
  reasons for, 187
  strategies for, 203–205
  theming performance, 160–163
  time tool, 201
  top tool, 201
  Valgrind tools suite, 202–203
testing application, building (example)
  creating application, 190–192

LSB Application Testkit (ATK), 196–197
overview, 189
testing with Mago, 193–195
themes
  defined, 147
  structure, 151–154
  theme engines, 148
  theme.xml file, 151–152
  tool for modifying, 150–151
theming
  objective of, 150
  overview, 147–150
  testing performance, 160–163
  Ubuntu MID, 148–150
  Ubuntu MID automatically, 156–157
  Ubuntu MID manually, 154–156
  Ubuntu Netbook Remix, 157–159
time. See also atime, disabling
  time tool, 201
  timelines (Clutter), 70
/tmp, size of, 208
toolbars, creating (Hello World), 62–64
toolbars, creating (Hildon), 59
toolkits. See also GTK (Gimp Toolkit)
  EFL, 79–81
  Glade, 66–70
  QT, 75–79
tools. See also pbuilder tool
  for application packaging, 105–108
  for automating image building, 226
  for benchmarking graphical operations, 156
  for caching packages, 225
  Linux performance testing tools, 201–203
  LSB Application Testkit, 196
  for modifying themes, 150–151
  Phoronix Test Suite, 197–200
  for testing, 197–203
top tool, 201
Torvalds, Linus, 7, 246–247, 255, 277
touchscreens, configuring, 214–217
trademarks, Ubuntu, 240–242
treb (Trebuchet) application, 131
tree object, (Git), 278
troubleshooting. See debugging; problems and solutions

U

Ubiquity graphical installer, 226, 239
Ubuntu
  Alternate Image, 226
  ARM, xxv
  Desktop Image, 226
  Developer Community, 245–246
  Hardy release, xxiii, xxiv
  Intrepid UMPC Project, 261
  Mark Shuttleworth and, 248
  Mobile and Embedded Project, xxiii
  mobile computing and, 254–255
  Mobile Developers team, joining, 262
  mobile project, xxiii
  Mobile Team, xxiv
  netbooks and, 6–8
  running on Arima, 261
  running on Freerunner, 260
  ubuntu-vm-builder tool, 30
  user interface, default behaviors, 224
Ubuntu MID, 54–55
  theming, 154–157
  theming and, 148–150
  Ubuntu kernel package, creating, 166–171
  Ubuntu kernel tree, updating, 175–181
Ubuntu mobile environment
  background, 11–12
  Jaunty Ubuntu MID release, downloading, 12
  KVM, 21–24
  networking in KVM/QEMU, 26–34
  networking in VirtualBox, 25–34
  QEMU, 21–24
  VirtualBox, 12–20
Ubuntu Mobile image example
  architectural support, 220
  build process, fine-tuning, 225–226
  building customized Ubuntu image, 230–235
  checking hardware, 220
  default Ubuntu image, creating, 226–230
  embedded systems and, 221–222
  Hildon default behavior, setting, 224
  important considerations, 219
  kernel fine-tuning, 221
  packages and repositories, 235–240
  policies/trademarks/copyright, 240–242
  power policies, defining, 221
  Ubuntu default behavior, setting, 224
  user interface, customizing, 222–224
  xfce default behavior, setting, 224
Ubuntu Netbook Remix
  germinating example, 234–235
  installing on UMPCs, 261
  Launcher, 54, 159
  overview, xxiv, 54–55
  poor performance and, 263
  theming and, 157–159
Ubuntu policy on packaging, 126–127
UDS (Ubuntu Developer Summit) for Intrepid, xxv
UMPCs (Ultra Mobile PCs), installing Netbook Remix on, 261
undervolting, 221
unique connection names (D-Bus), 309
universe package category (Ubuntu), 126
updating pbuilder environments, 124
uploading to PPA, 115–116
URI (Uniform Resource Identifier), defined, 189
USB (Universal Serial Bus)
  sticks, 257–260
  using, 260
USE_HILDON flags, 259–260
users
  business users, 129–131
  location-aware users, 138. *See also* GPS-enabled standalone web application (MID devices)
  multimedia users, 131–133
  social network users, 133–137
  user interface, customizing, 222–224
  usermode networking, 26
Usplash application, 158
uvccapture application, 220

V

Valgrind tools suite, 202–203
VDE virtual switch, 26
vertical boxes, GTK and, 66

319

Vesa graphics drivers, 262
video recorder (Entertainer), 132
virtual images, building custom, 31
VirtualBox
   installing and running, 12–20
   networking in, 25–34
   using bridge in, 28
virtualization
   CPU support for, 21
   defined, 11
"Visual Design of the GNOME Human Interface Guidelines", 150

W

Walters, Colin, 305
websites, for downloading
   dd for Windows, 260
   fix for Netbook Remix, 263
   GNU Haret, 259
   gOS 3 Gadgets, 134–138
   graphical germinate output for Ubuntu Netbook remix, 233
   Jaunty Ubuntu MID release, 12
   LSB Application Testkit, 196
   Marvell software, 252
   MID, old versions, 262
   PolicyKit (D-Bus), 90–91
   source code, xxvi
websites, for further information
   debuild command, 115
   DKMS, 181
   Netbook Remix, xxiv

plug computing, 265
port to armel architecture, 54
PPAs, 115
QT4 example, 75
REVU tool, 116
rules file (PPA), 109
Ubuntu Mobile commercial support, 219
Ubuntu mobile project, xxiii
Ubuntu policy document, 126
ubuntu-mobile release, xxv
"Visual Design of the GNOME Human Interface Guidelines", 150
webcam support, 220
Wheeler, David, 305
Wiener, Norbert, 244–245
window-specific settings (Hildon), 59
wrapper libraries (D-Bus), 305

X

X Window graphics adapter, testing, 162–163
x86 architecture, 250
XDG base directory specification, desktop, 57
xfce default behavior, setting, 224
xmag tool (images), 155
XSETTINGS (Linux), 156

Z

zenity suite (GTK widgets), 150
Zimmerman, Matt, xxiii